STURTEVANT AND DOBZHANSKY

TWO SCIENTISTS AT ODDS

STURTEVANT AND DOBZHANSKY

TWO SCIENTISTS AT ODDS

WITH A STUDENT'S RECOLLECTIONS

Edward Novitski
University of Oregon, Eugene, Oregon

With a foreword by James Crow
University of Wisconsin, Madison, Wisconsin

Library of Congress Number: 2004099656
ISBN :   Hardcover   1-4134-7084-X
               Softcover   1-4134-7083-1

Cover illustration by Edith Wallace, pinx.

This book was printed in the United States of America.

To order additional copies of this book, contact:
Xlibris Corporation
1-888-795-4274
www.Xlibris.com
Orders@Xlibris.com
25892

Contents

SECTION 2: THE EARLY YEARS

SECTION 3: HISTORICAL BACKGROUND

SECTION 4: CAREER AT CALTECH

SECTION 5: STURTEVANT AND DOBZHANSKY AT ODDS

SECTION 6: THE PLEASURE OF FINDING THINGS OUT

APPENDICES

REPRINTS OF RELEVANT PAPERS

This book is dedicated to my wife, Esther Ellen Rudkin Novitski; to our four children, Charles, Barbara-Jo, Paul, and Ellen; to our five grandchildren, Nancy, Linda, Sarah, Stu, and Elise; and to the many teachers, students, and colleagues I've worked with over the past 70 years who have valued the power of the intellect in their approach to scientific problems.

FOREWORD

JAMES F. CROW

DEPARTMENT OF GENETICS, UNIVERSITY OF WISCONSIN

A. H. Sturtevant was one of the brightest stars in the genetics firmament. As a student in T. H. Morgan's lab he had constructed the first gene map and later he pioneered in the study of chromosome behavior, as inferred from breeding experiments in the fruit fly, Drosophila. Th. Dobzhansky more than anyone else popularized neo-Darwinian evolution as an experimental subject. Educated in Russia, he came to the United States to tune up his Drosophila techniques and sought Sturtevant's tutelage. Together they opened up the field of using Drosophila populations to study evolution. Especially attention-getting was their determining the precise sequence of evolutionary change by the use of chromosomes with inverted gene sequences. They were close friends and Dobzhansky said that he owed his life to Sturtevant, who intervened in keeping him from being deported to Communist Russia. As a "Mendelist" he would not have survived long in his native land.

Then came the rift. To those who knew them both, this does not seem surprising. Dobzhansky was gregarious and flamboyant, worked intuitively and rapidly, covered a great deal of ground, and published relentlessly. Sturtevant was quiet and low-key, worked logically and slowly, did a few things very carefully, and published sparingly. It was inevitable that they would not persist as a working team. But how did the break come about? As far as I know, neither of them ever talked about it. Of course, there has been considerable speculation from others. Ed Novitski was a graduate student of Dobzhansky and Sturtevant in succession. Knowing both men and having the worm's eye intimacy of a graduate student puts him in a particularly good position to provide insights. And he does. Because of my own close acquaintance with both Sturtevant and Dobzhansky, I have followed the story with keen interest. I learned a great deal from reading this book. You will too.

Novitski has been the outstanding practitioner—a true virtuoso—in using genetic experiments in Drosophila to infer chromosome behavior. Several of his cleverest and most satisfying experiments are recounted here. An example was setting up elaborate matings, reminiscent of Joshua Lederberg's bacterial experiments, from which only the desired type of male was fertile. Then this needle in a haystack announced its presence by producing progeny in the midst of myriad failures. It is remarkable that Novitski stayed with this type of work, at a time when microbial genetics and DNA chemistry were grabbing the headlines. A second example is a long-time favorite of mine, a *tour de force*. He wanted to get a gene out of an inversion in which it was locked. This involved most of the tools in the genetic tool-kit. Ed obtained the equivalent of a double crossover by combining two single crossovers and eventually got these into a long inversion from which the gene could easily be extracted. He used triploids to get complementary aneuploids in the same gamete. And more, too complicated to explain here. Only Novitski would conjure up this kind of experiment and only he would carry it out. The book describes his inner satisfaction of doing this kind of genetics better than anyone else, despite the fact that most geneticists weren't interested. Like Feynman, Novitski experienced "the joy of figuring things out". In particular, as these examples illustrate, he employed the intellect—often in a flash of intuition—to reach a conclusion or result not arrived at by routine experiments.

Novitski also recounts his early years. He started experimenting with Drosophila as a high school student and continued throughout the rest of his career. He finished college in two years by taking exams in lieu of courses. He was accepted as a graduate student by, among other schools, the University of Texas. One year ahead of him and a graduate student at Texas, I heard from my classmates of an unusually bright and personable young man named Novitski, whom they met at a Genetics Society meeting. They very much hoped he would join us in Texas. To our regret, he accepted an assistantship at CalTech to work with Dobzhansky. He eventually became disillusioned with the Dobzhansky program, abruptly dropped his thesis project, and lost his teaching assistantship. Fortunately he was rescued by Sturtevant and began a lifetime of work in the Sturtevant tradition. It is clear where his sympathies lie in the Sturtevant-Dobzhansky rift. Nevertheless, he is, I believe, objective in his recounting of the facts.

The book includes a number of experiences, some most unusual. I won't summarize them, but will mention one observation that particularly struck me. Knowing that goats strongly dislike peppermint but will eat almost anything else, he suggested that a peppermint farmer should keep goats to clear out the weeds.

There is another side to Novitski. He has been Peck's bad boy of Drosophila genetics. He has a Feynman-like curiosity and love of pranks. These were sometimes complicated and elaborate, and always original. Several of them are recounted here. The clock example that opens the book is the most famous. It has enjoyed a wide circulation at meetings and in alcohol-stimulated conversation whenever two or more Drosophilists get together. Novitski has achieved what must be the highest honor in this realm: his reputation is such that he has frequently been credited with clever tricks that he did not perpetrate.

The four subjects of the book—the Dobzhansky-Sturtevant rift, chromosome mechanics, Novitski's education, and his pranks—are in separate sections. You can read any of them without reference to the others.

The book ends with several useful appendices, to which more technical aspects or background material are relegated. It includes a review by Sturtevant of a Dobzhansky book, which reveals a great deal about both men and a particularly insightful essay on Dobzhansky by his student Dick Lewontin. This essay gave me a new view and I found myself writing "how true" in the margins. Lewontin's point is that Dobzhansky was really a theorist, but lacked the necessary mathematical tools. His experiments were really *demonstrations* of an idea that he had already accepted theoretically. One of my own difficulties with Dobzhansky was that several times I could not repeat some of the data in his papers and I gradually came to distrust the details in his numerous publications. If these were really intended as demonstrations, it is not so surprising that his publications reported what he expected to find.

Finally, I would like to recount one of my earliest experiences with Novitski. In the early 1950s I was associate editor of Genetics. The journal had received a manuscript reporting that so-called "Minute" mutations enhance somatic crossing over. With one exception, as I recall, the strongest enhancement was on the same chromosome arm as the Minute. Novitski reviewed the manuscript and in his review he said that he had discovered that the one exception was caused by the mislocation of this particular Minute. It should have been on the same arm as the crossover enhancement. The hypothesis of the paper was strengthened and the author was greatly pleased. Sometimes a reviewer actually does improve a paper!

PREFACE

The organization of this account may be unusual. It is neither temporal nor logical, consisting of a series of sections that will be meaningful to persons with different backgrounds and interests. The sections have been arranged in order of (what is assumed to be) their interest to the average reader. It is hoped that no reader will find the book entirely incomprehensible, technical, or boring, although the latter part of the presentation will be of interest only to the professional biologist, or to the "Historian of Science."

The account is, at the outset, largely anecdotal, detailing a series of amusing or otherwise interesting events occurring during decades of work in the field of genetics. These could be palatable to any reader. This is followed by an autobiographical account of the early days of the author, which is presented to show the activities and thought processes of a young boy, with a logical-mathematical orientation, growing up during the depression years of the 1930s, and the sequence of events that led to a career in the field of genetics. During the period of graduate work which was started under Th. Dobzhansky and finished under A. H. Sturtevant, the writer became familiar with the personalities and work styles of both. Eventually they fell apart and engaged in an open confrontation. A personal view of this was initially the main thrust of the presentation.

The next segment concerns, an account of some of the unique triumphs and failures encountered in the pursuit of solutions to problems in this area followed by thoughts of the author and several others on the nature of research in science. Finally we have included the reprints of several articles pertinent to the overall presentation, including an unusually perceptive analysis by Prof. Richard Lewontin on the research style of Th. Dobzhansky and an article by Prof. A. H. Sturtevant commenting on several of Dobzhansky's scientific papers.

Many persons have read sections of this account and have reacted in different ways, ranging from complete negativity to unbridled enthusiasm. In about every case these comments have served as an incentive to improve or expand the effort to throw some light on the events of those early years. Some of those who have helped in this respect include: K. W. Cooper, J. F. Crow, J. R. Goodstein, M. Green, D. Leff, E. B. Lewis, D. Lindsley, R. C. Lewontin, E. Meyerowitz, W. B. Provine, R. J Robbins, and J. Yunis. Special thanks go to Prof. Raphael Falk and James F. Crow who have been unusually helpful in their critical remarks, made with an exceptional grasp of the issues involved. Particularly uninhibited and harsh criticism has come from members of the writer's own family, including Charles, Peggy, Barbara Jo, Paul, Nancy and Linda. Special thanks go to Elise whose keen eye and mastery of the language improved the text considerably.

This narrative was written initially for the benefit of inquisitive grandchildren, and therefore, from the technical point of view, includes some gross oversimplifications, usually by way of helpful analogies, when possible. On the other hand, criticisms by colleagues that some aspects were not being addressed adequately led to the introduction of more advanced concepts in this write-up. An attempt has been made to include both extremes by extended remarks, sometimes in the main text, but often by way of endnotes and, when more extensive, in appendices.

INTRODUCTION

Although envisioned initially as a casual account focusing on the problems encountered in the effort to acquire a Ph. D. in genetics in the late 1930s, the writing inevitably turned towards the relationship of two outstanding geneticists of those days, A. H. Sturtevant and Th. Dobzhansky, both of whom served sequentially as the writer's Ph. D. thesis advisor. Those two, although initially close collaborators, eventually became estranged. The author was involved to some degree in that separation.

Ordinarily one hesitates to discuss the relationships of others, particularly when there appears to be an element of animosity involved. In 1983, in a review of a book by E. A. Carlson, I wrote, in a discussion of H. J. Muller's relationship with other members of his laboratory group, "[We] regret that historians of biology find it necessary to embed the details of such personal conflicts in the historical record. We would rather remember them, and have future generations remember them, for the great contributions that they made to science and to humanity."

Some twenty years later I find myself contravening those pious words and doing precisely what I had condemned earlier. Briefly, it is in part because others (primarily historians of biology) with no first-hand knowledge of the events of those days have speculated freely, usually to the unwarranted detriment of one or the other of the participants. Second, the presentations have been inadequate and have skewed their view of research in genetics to conform to their own particular mind-sets. Third, because accounts of history are more likely to be written by those endowed with some prolificacy, the greater number of accounts favor one aspect over another and give the impression of a disproportionate contribution of one field over another. Finally, the overemphasis on personality rather than science has led to what might be called a cultish mentality on the part of the proponents of both individuals concerned. Nevertheless, full credit should be awarded to those who, without any association with the two scientists involved, either within the laboratory or outside, have tried to shed some light on their relationships. These efforts are usually couched in the most elegant literary output and, where appropriate, in highly imaginative and creative formulations. To avoid undue petty controversy, references to such endeavors, particularly when they are at odds with my own experiences and perceptions, are minimized.

Much of the confusion surrounding the complicated personal relationships of those two well-established geneticists may be clarified by focusing on their intelligences. In fact, the biases of the author and even those of the Historians of Science can be better understood when viewed in the light of "multiple intelligences."

During the past decade, some psychologists have proposed that "intelligence" is not a unitary property of the individual, but rather consists of about eight or nine discrete, and more or less independent, aspects of ability (Gardner 1999). Those that are of most concern to the discussions found here include the linguistic, the logical-mathematical and the naturalistic. The first is found to a high degree in writers, poets, speakers, lawyers and historians of science; the second in mathematicians and physical scientists, and the third in biologists.

A good illustration of the first two differences can be seen in a comparison of Mendel (1864) and Darwin (1868). The brilliant exposition of the laws of inheritance using the garden pea proved that the first was obviously superb in the logical-mathematical area but his writings were brief and to the point.

Darwin, on the other hand, excelled in the linguistic aspect; his writings are voluminous almost to the point of being boring. Yet when he performed a simple cross of two forms of snapdragon, he noted that one form did not appear in the first generation, and when he got an approximate three to one ratio in the second generation he did not ask even the most obvious questions, attributed the results to "gemmules," and failed to make those obvious tests that Mendel did.

Thus, highly talented persons, as both Sturtevant and Dobzhansky were, can differ markedly in the degree to which they possess different kinds of intelligences. In some instances these intelligences, different in two persons, may complement each other for a decidedly beneficial interaction. In other cases, they may conflict to produce a lack of understanding, even mutual antagonism. We shall see both of these in operation in the relations of Sturtevant and Dobzhansky.

The writer must confess at the outset to a bias in a certain direction. This comes about, in part, because the distribution of his intelligences, and particularly his deficiencies, cause him to be more sympathetic to the logical-mathematical point of view than to the naturalistic and linguistic ones. If this account differs appreciably from those presented by others, it should be considered that they also may have a view biased by their own particular set of intelligences, of which the linguistic may predominate.

SECTION 1

FUN AND GAMES

Most of those unfamiliar with scientists or their work visualize their activities as a sober, dignified, even monotonous routine of life over a microscope, or peering at a test tube held up to the light, and to some extent this is true. But occasionally the sober routine of scientific endeavor is interrupted by events or situations, mostly promoted by a spirit of good-natured fun, which may provoke a smile. Further, while it is true that the main effort of most in science must be directed towards a single area, it is also the case that many have rather wide ranging interests in different areas of science, usually unknown to their colleagues. Here are a few examples:

Seminar in Seattle

Shortly after arriving in Eugene from Oak Ridge in 1958, I received a letter from Herschel Roman inviting me to Seattle to give a seminar. He and I were old friends from our days at Caltech together, and I replied, playfully, I thought, that I was glad to know that the people at the University of Washington were upgrading their genetics program by inviting competent workers to teach them about genetics, and that my seminar in Seattle would give the group there a much needed lift, etc., etc.

I finally got to Seattle and in his introduction of me to the group, Roman read the letter I had written. As my face got redder and the audience roared in amusement, several thoughts crossed my mind, and, as best as I can recall, they went Snoopylike something like this: "Alright for you, Herschel Roman, I'll get even with you some day!"

And get even we did, beyond all expectations. I invited him to Eugene the following year to give a seminar, and made elaborate preparations to do him in. We arranged for one of our more vocal graduate students to sit in the front row and to ask, every three or four minutes, the most stupid questions he could think of. One of the lab assistants took along a pile of books and loose papers sat at the end of a closed row so that she could not leave without completely disrupting the seminar and about ten minutes into the seminar she noisily said "Pardon me" about ten times, finally got to the center aisle, then to leave went down towards the speaker where she "accidentally" dropped everything at his feet. As he bent over and helped her recover her stuff, she remarked very loudly that she had come to the wrong class and that she did not like this subject at all. Roman later interrupted his talk with the comment that this was the most unusual audience he had ever spoken to!

But the jewel of the performances was the manipulation of the clock. These seminars were held in a physics classroom, which had next to it a prep room with an elaborate switch-board through which went all power lines. In preparation for this seminar, we had gotten an audio oscillator, determined where the clock was hooked in, and arranged to feed in a frequency of our own choosing. After some discussion, the group decided that a slow clock would cause him more dis-comfort than a fast one.

It was necessary to divest Roman of a wristwatch, if he was wearing one, so that he could not know the real time. So, a few minutes before four o'clock, John Erickson appeared, with an urgent request; "I am starting a timed X-ray experiment and I forgot my watch. Can I borrow one from someone?" Roman graciously obliged and the stage was set.

Before the seminar started I called his attention to the wall clock, which now read 4 p.m. and told him that we liked to have seminars start on time and finish precisely at 5 p.m., to which he casually agreed. Despite the rude interruptions, he finally finished and when he looked at the wall clock, it read 4:40, twenty minutes still to go. I remember his puzzled re-mark "I've given this talk several times before, and never have I finished this early." An awkward pause followed and then he said: "Oh yes, and one more thing" and he continued for an-other five minutes on a subject that he clearly had not initially planned on including in his talk. In the meantime, five o'clock (real time) came and about half the audience, those with other responsibilities at that time, got up and left. Some of these were good friends of Roman's and this was the crowning blow. He was (in my opinion) devastated at having given such an obviously unacceptable performance (in his opinion) that the audience were walking out! I almost felt sorry for him.

I had to call a halt to the seminar before the wall clock said 5 p.m., and then told him how we had deliberately done him in. The group responsible got together with him and recalled in vivid detail all the things we had done to make his one hour in Eugene memorable. He laughed with us, was a good sport about it, and our friendly relationship never changed.

The next year the chairman of the seminar committee at the University of Washington wrote asking if I would please come and give a talk. I politely declined.

Several weeks later, the cytologist Andrew Bajer returned from Sweden where he had spent time with the well-known cytology group at Lund University, and when I asked him if anything memorable had happened during his stay, he said no, but that he had heard of a hilarious incident involving a clock in Eugene, Oregon! Apparently even before the internet interesting news could travel great distances quickly.

Years later I read, in one of the books describing pranks of the students at Caltech, an account of how the students befuddled a faculty member by monkeying with a clock. This is such an obvious trick that it probably was original with them. Nevertheless I cannot help wondering if the idea did not originate somewhere in the Great Northwest and slowly make its way over the Siskyou Mountains into Southern California.

Several years ago Dan Lindsley reminded me that altering a clock might not have been a completely original idea. Back about 1950 when the three of us were together at the University of Missouri, Alexander Fabergé persuaded a clock in our lab to run backwards as a way of confounding a neighboring apparently lonely physiologist who seemed to come into our lab too frequently to "see what time it is." I myself became painfully aware that synchronous motors always come in two forms, one running clockwise and the other running counterclockwise, when I replaced the tracking motor on a reflecting telescope.

Stern's Brilliant Reply

Playful, humorous or even slightly derogatory introductions at seminars are possible when the subject is well known to the introducer, is of the same age group, and can be counted on to take such an introduction in stride or even to reply with a snappy comeback. One would never try such an introduction to an older well-established person (Sturtevant, for instance), where any such attempt would be immediately sensed as a gross impertinence.

However, one exception I made to this rule was the case of Curt Stern. Stern I knew very well. I spent two years with him at the University of Rochester right after World War II, one year during the tenure of my first Guggenheim Fellowship and a second year working with him on an Atomic Energy Project. Our relationship was closer than I had experienced with any of the others in the older generation that I worked with before or afterwards. He was the only person to whom I confided, after twenty five years, that when I was cleaning up Calvin Bridges' lab after his death, I had pinched his file of photographs.

From his side, he was our first baby sitter, taking care of newborn Charles, after insisting that Esther and I deserved a night on the town on our own (which meant going to a movie!). Esther typed the first edition of his book on Human Genetics, transcribing from his almost illegible scrawl, and my first introduction to that field came from my proofreading.

When I asked Stern to come from Berkeley to Eugene to give a talk, he replied with a short note written on a small piece of paper, the first side of which ended with "The Action of the Y-Chromosome in Sex Determination" followed by P. T. O. (which to me clearly meant Please Turn Over the page for the rest of the letter).

I gave this letter to the secretary responsible for making the signs that publicized seminars and after his name and affiliation she put into big letters "The Action of the Y-Chromosome in Sex Determination P. T. O." The signs went up at the usual spots and when I saw them my first inclination was to pull them and put up corrected ones. But then my better judgment prevailed.

When I introduced Stern, I gave the usual accounting of his many successes in genetics and then ended with: And now Prof. Stern will give his talk on "The Action of the Y-Chromosome in Sex Determination P. T. O." When he heard my P T O addition his brow wrinkled. Of course he had seen the notices with the mistake, but apparently thought it was harmless. Now it had come back to challenge him.

As he talked about the Y-chromosomes in different species, his face relaxed, he looked at me and smiled and I knew that he had somehow found a way out. He finished his seminar (which everyone agreed was excellent) and then went on to say: "For those of you who don't know what the Y-chromosome P T O means, it means The Y-chromosome, the Primary Testicular Organizer!"

I think that only he and I knew that he had bested me in a little game of wits. Hoist with mine own petard, to paraphrase Hamlet.

Squirrel in Columbia

Sometimes even the mildest of jokes can take an unexpected twist. One such case happened when a graduate student (whom we shall call Walt) at the University of Missouri captured a squirrel that somehow had wandered into the building and made its way down to our lab. Walt found a cage for a home and quickly bonded with the squirrel with the same affection that a mother feels for a newborn child. It was his baby and nobody could say otherwise.

But it soon became apparent that the creature was not thriving. Not only that—its captivity clearly was killing it. We all (except Walt) realized it had to be released. But how to do it without upsetting Walt?

Finally somebody got a bright idea. It could have been me, and I would like to take credit for it, but it might just as well have been one of the other lab workers. I don't remember. The plan was to go up to the attic of LeFevre Hall and see if there was a squirrel in the vast collection of animals the University of Missouri had inherited when the St. Louis Exposition closed down in 1904, some fifty years earlier. That collection was enormous, its existence being revealed by the presence of a stuffed gorilla at the top of the stairs on the second floor, an animal who regularly had a cigarette inserted between its lips by some student who thought he was playing a joke never thought of by any other student and just as regularly had the cigarette removed each morning by Cook (short for Noble Cook Mitchell), our patient janitor.

Sure enough, there was a squirrel in the attic, so, in Walt's absence, we released the live squirrel to the outside world and replaced it with the stuffed one. Within a few hours Walt showed up at the lab and went immediately to the cage. He reappeared shortly with the comment "You know, I think our squirrel has had it." So far so good. But what happened next was not part of the plan. Walt continued: "It must have had some kind of disease. I'm going to take it over to the Veterinary School and see if they will do an autopsy on it." At that point more considerate sensitive souls would have restrained him somehow, but we just sat back a bit aghast at this turn of events. Walt carefully wrapped up the stuffed squirrel and disappeared out the door.

We can only imagine what happened next, with the people over at the veterinary school being asked to do an autopsy on a stuffed squirrel, at the same time as Walt was insisting that the animal had just died. Anyway, he returned to say that he thought that the people over there were a bunch of incompetents who couldn't tell a squirrel just deceased from one a stuffed one!

I have no doubt that eventually somebody confessed to him that it was a hoax. For my part, I still have a feeling of guilt that I never restrained Walt from taking the stuffed squirrel to the veterinary school for an autopsy.

Blast of the H-Bomb

I spent a couple of years at the Oak Ridge National Laboratory in Tennessee in the late 1950s where my love of gadgetry found an outlet in their high speed computer. Named the ORACLE (for Oak Ridge Automatic Computing and Logical Engine!), it was vacuum tube powered, had a cathode ray tube memory, and occupied several rooms with plenty of air conditioning. To use it, it was necessary to master the language of the machine, and its limited memory required the skillful manipulation of arrays of data operated on by sets of numerical arithmetical orders.

This was a challenge, and I think that I managed pretty well (see Appendix F). At the beginning I was tutored by a very competent woman who had come from the Aberdeen Proving Grounds where the government had an earlier computer (the ENIAC, if I remember correctly). She told me the following story:

At the time of the design and construction of the H-bomb, a team of which she was a member were given the job of using the computer to solve a large number of equations necessary to make the bomb work efficiently. They had no problem setting up the series of necessary machine orders and after some time they were all set to go. But when they ran the program, it came to a screaming halt, unable to proceed. After some investigation, they learned that the reason for the halt was that the machine was being asked to take the square root of a negative number, which the machine recognized as an impossibility. So the group went into the debugging mode, which, with machine language of those days, was no easy task. Days went by, and after repeated efforts nobody could find anything wrong with the program. Finally somebody called from Washington to say that the construction was being seriously delayed and that they needed the results immediately.

In desperation, they decided that the only solution was to insert another order immediately preceding the one giving trouble which, in effect, said "if this number is negative, make it positive." Then the program worked fine, they got their numerical results and transmitted them upwards.

She said that when the H-bomb blasts went off at Einewetok, and the reports were that the bombs had a bit more wallop than was expected, she was sure she knew why. But mum's the word.

Whether this is true or she was pulling my leg I don't know. But it makes a good story.

Frank Stahl Gets the Nobel Prize

One morning the University of Oregon student newspaper, the Daily Emerald, carried in a headline the great news that our brilliant colleague at the University of Oregon, Frank Stahl, was one of three who had received the Nobel Prize. Frank was one of those persons who can be described as a true genius, and if anybody deserved this recognition, he did.

However, on closer inspection, it appeared that in fact Watson and Crick had been so honored, and the article simply stated that Stahl was doing work in the same field. Apparently the student writing the headline hadn't bothered to read the article. Anyway, this opened an opportunity too luscious to pass by.

After some cogitation, the Drosophila group decided to see if we could shake his usual calm composure. One of the female students, a little more aggressive than most, undertook the role of a reporter for the Associated Press and would pretend to call him from New York (although in fact she would be calling within the University system, and not more than a hundred feet or so away.)

I can recall the gist of the telephone conversation at our end but at the other end all we could hear was a series of explosive sounds which I am indicating by the word "no." It went something like this:

Associated Press: Good morning, Professor Stahl. We at the Associated Press here in New York have heard the good news that you have won the Nobel Prize in Medicine.

Stahl: No No NO
AP: We would like to get more information about the work
          you did . . .
Stahl: No No NO
AP: We understand that you scientists are always very mod-
          est about your accomplishments and are reluctant to
          discuss your great discoveries.
Stahl: No No NO
AP: Of course if you are too modest I can get the informa-
          tion elsewhere.
Stahl: No No No NO
AP: Congratulations again Professor Stahl, and you can be
          sure that I will get this interview on the AP wires im-
          mediately.

And at that she hung up. We can only guess what Stahl's next hour was like. No doubt it included a frantic effort to get in touch with AP headquarters to tell the puzzled person at that end that he had not won the Nobel Prize! I do not know what happened. I have never discussed his matter with him and he may still be unaware of the source of the joke, although it is difficult to keep incidents like this quiet for any length of time.

I do recall that Stahl was heard to remark later that the greatest indignity was not the unfortunate headline so much as the fact that the editors of the Daily Emerald (who must have had a keen sense of the relative importance of news stories) had chosen to put this article on the third page!

Churchillian Humor

At the beginning of WWII, the press reported that Churchill objected to a correction a secretary made to one of his written speeches, repositioning a preposition which he had used to end a sentence with. Churchill wrote in the margin where the sentence had been corrected, "This is the sort of nonsense up with which I will not put."

Sturtevant, my thesis supervisor at the time, stuck his head in my doorway one morning as he was on his way to his office-lab, told me that story and then went on to relate his own favorite sentence, which ended not with one but with five prepositions! It seems that a youngster was objecting to being taken up to bed at night to be read to by his father. He said: "Daddy, what are you taking the book that I don't want to be read *to out of up for*?"

After several days I triumphantly reported back to Sturtevant that I had improved his sentence. Now it ended with eight prepositions. Since the book was about Australia (I said) the sentence should read: "Daddy, what are you bringing the book that I don't want to be read *to out of about down under up for*?"

I had cheated a bit, obviously, but he clearly enjoyed the one-upsmanship involved.

Galloping Gertie

The director of the Fulbright program for the Netherlands was a super-efficient woman named Wobbina Kwast. She had the bright idea that at a gathering of Fulbright Fellows and their wives in Amsterdam we should form a circle, then peel off by couples, proceed to introduce ourselves to each of the others in a clockwise direction, engaging in a brief conversation with each of the other pairs. To start things off, she quickly mentioned the names and the specialties of each of the Fellows. I got a bit worried. Physicists and chemists I can talk with more or less intelligently, but a civil engineer? As I made the rounds I pondered the problem I would be facing shortly.

Then I had an inspiration. What happened when I confronted our civil engineer is shrouded in the dismal recollections of the people there. What I remember is that when I approached him I said "Good evening, sir. I wonder if you can tell me more about why the Tacoma Narrows Bridge collapsed?" Esther's recollection is somewhat different. She remembered a remark more like this: "Hey, do you know who the idiot was who designed Galloping Gertie?"

Whatever the introductory comments were, I realized I had made a big mistake when his wife grinned broadly and poked her right forefinger in his direction. It turned out that he was, in fact, one of those involved in the design and later observations on this structure whose gyrations under high winds eventually caused it to crash to the water below. It seems that the engineers were aware of this propensity and had constructed protective shields to divert any excessive air flow, but had underestimated the disastrous effects of unexpectedly high winds.

My natural tact inhibited me from asking if he was involved in the design and construction of the replacement that eventually went up.

Sic Transit Gloria

During the train ride to Pasadena at the start of my graduate career, Dobzhansky told me several stories worth recalling. One of these involved a very well-known cytologist named Karl Belar. He had done some of the pioneering work on cell division using material from salamanders and lilies and came from the Kaiser Wilhelm Institute for Biology in Berlin-Dahlem. His career as an addition to the department at Caltech was cut short by a fatal car accident in the desert, while on a field trip.

Belar's remains were cremated. His wife had no interest in acquiring his ashes and they became the responsibility of the biology department. When I arrived to start my graduate work I ran into his name in my reading and my curiosity was aroused. What had happened to the ashes of this distinguished scientist?

Dobzhansky told me that they had been put into a large jar, which was stored on a shelf in a first floor laboratory used by the elementary biology classes. I went to that room and made a thorough search but could find no trace of his ashes, so I asked the janitor, Mr. Bonas, if he had seen them.

"No," he replied. But then after a pause he said "Last summer I cleaned out the lab and there was a large jar with some stuff in it."

"What happened to it?" I asked.

"Oh," he said "I flushed the stuff down the toilet and cleaned the jar out."

I went back to the beginning biology lab and, sure enough, there on the top shelf was a jar of the right size, sparklingly clean.

It occurred to me that someone should have put a label on that jar.

An Educational trip from San Diego to Ensenada

Dan Lindsley's car was waiting to take a hastily assembled group of us to Mexico. At the last minute I clambered aboard and took the only remaining seat. A trip from San Diego to Ensenada, Mexico, can be completely boring. I have always found the monotony of the passing desert landscape conducive to sound sleep. However, this trip would be different. The seat adjacent to mine was occupied by a girl, clearly an extravert with highly social tendencies, and in a short time we were deep in conversation. She was interested in my background and when, after some questioning by her, I mentioned my activities in human genetics, she became more engrossed.

To start with, I explained to her the chromosomal basis for some of the more common human anomalies, although I had to repeat my explanations in several different ways before she caught on. Her slowness was undoubtedly caused in part by her foreign background because, although she spoke English fairly well, she had an unmistakable foreign accent. For my part I spoke loudly and clearly to help her understand. This I could do easily because I had had practice with the same material before classes of a hundred or more students.

All in all, I think that I did fairly well. For several years previously I had attended the Genetics Clinic at the University of Oregon Medical School, a distance from Eugene of about 110 miles. This took abut two hours of travel time each way the first Wednesday of every month. Thus I was familiar with the most common genetic conditions of medical interest. I had brought together my lecture notes for an elementary course in human genetics and put them into a textbook which reversed the traditional sequence of presentation of providing the basic biological phenomena first. I had found that students, particularly those majoring in other disciplines, could be profoundly bored by the detailed description of cell division at the beginning of the course. Instead, I considered a series of human conditions first, and then provided the biological basis for those conditions as I discussed each. So I was well prepared to explain the elementary principles, with specific examples. As I tried to make clear the fundamentals of human genetics to my traveling companion, I emphasized the role of the X and the Y chromosomes in human development, and the developmental effects of additions and subtractions of those chromosomes from the normal chromosome complement.

After about three quarters of an hour I had finished my exposition just as we arrived at our destination. We all got out of the car and, as we were stretching our legs a bit, I looked more closely at my girl companion. She was in fact a bit older than I had thought, a young lady, rather attractive and quite composed. Dan Lindsley, who had been the driver of the car and also no doubt the recipient of my lecture on the principles of human genetics, came over and said:

"Ed, I don't think you have met Pat Jacobs."

I gasped. No, I hadn't met her. But I knew her name very well. She was unquestionably the most competent and best known worker in the field of human chromosome anomalies, and her pioneering work made her preeminent in the field of human cytology, not just in England, but worldwide. I had to refer to her scientific papers repeatedly as I worked on my human genetics lectures, and later on the textbook. And I had just given her a short course in human genetics! I did not ask her how well I did but I am sure that I made any number of errors in my exposition and that she must have concluded that I was a complete idiot.

Clearly the group had singled me out as particularly naive, prone to fall easily for a simple practical joke and had stage-managed the whole affair. Of course everybody thought that this was a hilarious performance and laughed accordingly. Whether I did too I do not remember.

But I do think that it is regrettable that anyone would subject a close friend to such humiliation and embarrassment, something that I, having a more sensitive nature, would never even consider.

Our Plane Catches Fire

Flames were pouring out of the engine on the right wing and I had a perfect view. Before I had a chance to become apprehensive, the stewardess dashed down the aisle and sat next to me, where she too had a good view. She was visibly disturbed, apparently more concerned than the situation warranted.

"The pilot asked me to watch and see if anything more is happening," she explained.

"Well," said I in as reassuring a tone as I could manage. "I flew many bombing missions over Germany during World War Two and have seen this sort of thing many times. It's nothing to worry about."

This of course was an outright lie. It is true that I went on about a dozen bombing missions. However, as a radar officer I was considered nonflying personnel and could go only as an invited guest on nonhazardous missions, known as "milk runs," towards the end of the war. On these flights the anti-aircraft bursts were sporadic and far off the mark; I think that the German gunners, sensing the inevitability of defeat, were deliberately aiming away from our aircraft. Anyway, I not once found myself confronting what might have been a life-threatening situation during the war as I now did on this short hop from Knoxville, Tennessee to Cincinnati, Ohio.

"What do you think is wrong?" the stewardess asked.

"Well," said I, trying to assume an air of superiority, "It's probably nothing more than a broken piston rod."

This did not seem to lessen her obvious discomfort, which surprised me. After all, I knew from movies I had seen that passengers were supposed to cower in fear, while stewardesses always walked up and down the aisles reassuring them. Here it was reversed. In an inspiration designed to lessen her discomfort I said:

"Would you care to hold my hand?"

This startled her like a quick slap across her face and she replied brusquely "Certainly not," got up, and disappeared into the pilot's cabin.

Anyway, we made an emergency landing at a small field in Kentucky and finished the leg of the journey to Eugene, Oregon the next morning.

After a few days in Eugene I made the return trip and, as luck would have it, in Cincinnati I was greeted at the door of the plane by the same stewardess. We grinned at each other like old friends, as was appropriate for two who had survived a life-threatening experience together. "Well," said I, "Did they find out what was wrong with the engine?" "Oh," said she "I thought you knew. It was a broken piston rod."

Since then I have from time to time wondered whether prop-jet engines really have piston rods.

Eisenhower's Economy

Do papers ever get published when they are based on false data? I myself was involved in one such a case that to me, at least, had a humorous aspect.

It was known for a long time that there are more males born than females in the human population. Since the process of meiosis should result in equal numbers of X and Y bearing sperm, which would then produce the two sexes initially with equal frequencies, it was natural to assume that this deviation from 50-50 is the result of some change in the internal environment of the mother as the embryo develops, with this change affecting one sex more than the other. It was also well known that the excess of male births decreases as the mother grows older.

Some work we did with Drosophila suggested that the male parent also might have an effect in changing the ratio of the two sexes produced. So we went to library and discovered that the Yearbooks of Vital Statistics had just started including tables that gave the ages of the parents, along with the sex of the newborn simultaneously, making it possible to do the appropriate statistics.

Much to our surprise, the work with these limited data indicated that the shift in the ratio of boys to girls (usually given as the proportion of males births over the total) that occurred with increasing ages of parents was highly correlated with the increasing age of the father, and not at all with the age of the mother (Novitski 1953). This was such an unexpected result that it evoked considerable skepticism (see, for instance, the criticism by M. Bernstein, and the response Novitski 1954) and to this day is viewed with suspicion by some workers!

Nevertheless, the results needed support. This was provided in a series of subsequent papers (Novitski and Sandler 1956, and Moran, E. Novitski and C. E. Novitski 1969).

In 1957 I visited the Bureau of Vital Statistics to see if I could get more detailed data on human births. After the customary formalities of introductions and pleasantries, one of the two statisticians in charge said:

"Oh, yes. You are the one who has been using our 50% statistics!"

"Really," said I. "What do you mean?"

"Oh. Didn't you know? Those tables [in Novitski and Sandler 1956] which included the years 1951 to 1953 were made up by transferring the birth data, as we received them from the states, to punched cards from only half the birth records and then multiplying the resulting totals by two."

It appears that the head of the Department of Health, Education and Welfare, Oveta Culp Hobby, was following orders from President Eisenhower to economize and so they cut the cost of the table-making in half. Why they didn't cut the costs even more by recording only one out of every ten births, and then multiplying the result by ten puzzles me; this could make sense to a mathematically and statistically challenged person, as the politicians seemed to be. Our error developed in the following way:

In order to combine the figures from the newer tables with those from the previously analyzed ones, we used the services of a graduate student in ecology who got paid for the work by the federal government in a work-study program. This student added to our old totals several hundred numbers from the newer tables apparently without once realizing that there was something peculiar about the numbers, that every number in the newer tables always ended in 0, 2, 4, 6, or 8 and never in 1, 3, 5, 7, or 9! I would say that that student was arithmetically challenged. And of course it was basically my responsibility for not monitoring his work more closely.

Anyway, we eventually made a public acknowledgment of our error in a subsequent paper (Novitski and Kimball 1958), pointing out that even though our estimates of the error involved in the work were slightly incorrect, the overall conclusions remained valid.

However, the most conclusive analysis was provided by Ruder (1985), who, using magnetic tapes I got from the Bureau of Vital statistics covering over two million births and more than a dozen variables, showed, by advanced statistical procedures and the precision afforded by the high-speed computers then available, that the age of the father was indeed significantly correlated with the change in sex ratio, with the age of the mother being relatively insignificant.

Bureaucratic Complications

Usually bureaucrats who safeguard the public purse are overbearing and astonishingly stupid in enforcing rules and regulations, and sometimes they are not. Here's an example of each:

A fairly distinguished guest from England was visiting the US, and we invited him to spend a few days in Eugene (with, of course, the stipulation that he give a talk at the University). He agreed, fulfilled his commitment admirably, and dutifully filled out a form that would enable us to compensate him for his expenses while here in Eugene. I sent the information to the business office, which transmitted it to Salem, where some clerk in the state finance business office spotted an unforgivable transgression. Our guest had included, along with his hotel room and meals, an impermissible item, the cost of a glass of wine before dinner. Presumably this trivial problem could have been handled by simply striking out the offensive item, but instead the office feared the repercussions from the electorate if it ever became known that an expense account with liquor mentioned crossed the auditor's desk. So they bounced it, with the demand that a new account be prepared, this time eliminating any mention of the demon rum. I rather embarrassedly sent this paperwork on, to catch up with him in England. Obviously irritated, he replied to me that he expected to be treated somewhat better than that when he visited the colonies and that he would pay the bills himself rather than be the subject of such indignities.

An opposite attitude was that of the head of the university business office, who seemed to have the unorthodox view that his function was to provide a service to the faculty and student body. Let me tell you how he solved a problem for me.

During my first sabbaticals abroad, to Switzerland and to Australia, I found an Oregon State rule particularly annoying. My fellowship funds always included a small amount for travel and these funds were to be administered by the university, but the state rules said that whenever a person traveled out of the state of Oregon, permission at the highest level was required. I guess this became a rule when some governor, finding that he had some time on his hands, decided to satisfy his curiosity and see how state employees were having fun traveling while he was stuck in his office in Salem. This was before the days of fax machines and such gadgetry, so we depended on the mail system, which was abysmally slow. Even airmail required several weeks for a confirmation. This made acquiring approval for short, quick trips almost impossible.

Anyway, I had another sabbatical coming up, was planning to do something daring (planning to go to the Netherlands) and could foresee difficulties in getting scholarship funds for short trips within Europe. So I visited our business manager, a Mr. McGilllicuddy, whom we called Mac.

"Mac," I said, "I have a problem. This rule about getting prior permission for out-of-state travel is a real pain. It just takes too long to get prior permission. Is there anything that we can do about it?"

Mac looked reflective for a moment and then he said: "I think that I have an answer. Listen closely. What's the difference between in-state travel and out-of-state travel? If you go from Eugene to Salem do you cross the Oregon border?"

"No."

"That's in-state travel. Now if you go from Eugene to Seattle, do you cross the Oregon border?"

"Yes."

That's out-of state travel. Now if, when in Holland, you go from Amsterdam to Paris, do you cross the Oregon state border?"

"No."

"Then that must be in-state travel."

Thus it was that I felt free to travel on field trips, go on various seminar trips, etc. in Europe without the required prior permission for out-of-state travel!

Surfing in Hawaii

Our son Charles accompanied us to Australia (as did our other son Paul, and the younger of two daughters, Ellen). There Charles met an attractive native and, after a rather brief courtship, the two got married. It was in the middle of winter, on June 15, so the two celebrated their nuptials by skiing in the Great Snowy mountains, where, that year, as usual, there were two inches of snow on the ground. Then they went off on a honeymoon trip to Japan before heading back to the US, where Charles would resume his undergraduate career at Columbia University.

I was also headed back to the States and Charles and Peggy and I arranged to meet and spend a few days together in Hawaii, where we would take advantage of the beaches and other attractions that that island had to offer. In addition, a symposium on the use of computers in the solution of genetics problems was being held there at that time and I would present a short paper on the population dynamics of self-sterility genes in plants.

One afternoon Peggy and I made it into the water first, and were splashing and paddling in shallow water, when we were approached by an old friend, Herman Lewis, who at the time was the program director for genetics at the National Science Foundation. After the customary greetings I said (with tongue in cheek):

"Herman, I would like you to meet the new Mrs. Novitski." Herman's eyes opened wide. "When did this happen?"

"The ceremony was on June 15," was my reply.

He paused for a bit, then came the next question: "What happened to Esther?"

"Oh, she's doing just fine. She's over there on the beach."

And I waved my arm vaguely towards the mass of people on the beach. Actually she was not on the beach at that time. She had gone ahead to Oregon to take care of some family business.

Herman glanced at Peggy, got a worried look on his face, and, after a brief pause, paddled off away from us. Undoubtedly he was preoccupied with his conclusion that he had inadvertently stumbled upon a highly personal situation about which the less said the better.

I guess I saw him again in Hawaii; he was attending the same computer conference I was, but I do not know when, if ever, he became aware that my answers, while strictly speaking, were quite correct, he had simply jumped to the wrong conclusions. Apparently he did not allow his perception of my Bohemian lifestyle warp his good judgment because my National Science Foundation Research grant was routinely approved by his committees over the next few years.

Adventures in Languages

Whenever Esther and I spent a year or more in a foreign country, we made an effort to master the native tongue (although we made exceptions to this in Finland and Australia). Holland proved to be particularly challenging. The guttural "g" is more than a non-native can manage, the common ac-

knowledgment of this being the usual non-Dutch pronunciation of the name "Gogh" as simply "go."

A group of us at the University of Leiden hired a tutor to
teach us the elements of the Dutch language. She was from a
southern province, with a distinctive dialect, in which the guttural "g" could be best described as gentle as the purr of a
kitten. Try as she might, she could not transmit this art to any
of us. Nevertheless, we did learn a bit and I can recall with
some pride that after a few months of lessons, when I was asked
by a Dutch passerby on the street what time it was, I could, with
the greatest nonchalance, point to a large clock half a block
away.

But a little knowledge can be a dangerous thing, as I was
about to discover. One afternoon one of the graduate students
gave a seminar about a research project that involved looking
at the chromosomes of individuals with a rather uncommon
defect, something called "fragile X-chromosome," a defective
chromosome which has its major manifestation in the male
sex. Although the seminar was given in English, one Dutch
word appeared frequently, the name of a little village between
Leiden and the North Sea coast.

The road connecting Leiden and the small coastal town
where we lived, Noordwijk-an-See, went through the smaller
town of Vorhout, so I knew it well. This town seemed to be the
origin of the material mentioned by this student in her presentation. I was intrigued by the possibility that such a small
village should be the origin of so many cases of this anomaly.

When the time for questions came at the end, I asked a
simple one: "How is it that so many of your cases seem to come
from the tiny village of Vorhout?" The student at first seemed
puzzled by my question, and then her face turned a brilliant
red. For a moment we were at an impasse, which was interrupted by Bert van Zeeland: "Oh," he said, "You are confusing
Vorhout, the town, with Voorhuid, which means foreskin."
These two words, although spelled differently, sound exactly
the same to my non-Dutch ear. It seems that her material was

coming from the foreskin removed from newborn males at the time of circumcision, a circumstance which would not be expanded on in proper Dutch society. Inadvertently I had crossed the line that separates good taste from vulgarity.

The student escaped from this awkward moment and fled the room. I saw her occasionally in the corridor during the next week or two, but she would always quickly turn and move away, obviously not wanting to be reminded of the embarrassment I had caused her. And then she disappeared altogether. I was told that she had decided to change her field of interest and was working in another department. I have sometimes wondered whether it was my incompetence with the subtleties of pronunciation in the Dutch language that was responsible for removing a promising graduate student from her chosen field of endeavor.

In more recent years, a different question has come up. What is happening to those millions upon millions of tissue samples, all perfect for DNA identification, that are regularly being produced at the time that parents either insist upon having, or agree to have, these tissues removed from their newborn males? My thought is that, considering the emphasis being placed upon this sort of identification, the federal government is in fact storing these away for future use in identifying terrorists, criminals, dissidents, nonconformists, and other unpleasant people. And this huge data bank would go unnoticed if the depository were not in the United States, but in Europe, in a little Dutch village appropriately named Vorhout.

DuBridge and Hot Microwaves

There are those rare occasions in life when several apparently independent events coincide to produce a burst of good fortune. This seemed to be the case with my choice of the University of Rochester as a school to spend a postdoctoral year as a Guggenheim Fellow on the one hand and the presence there of Prof. Lee DuBridge, at that time Head of the Physics Department (and later President of the California Institute of Technology) on the other.

During the last year of World War II, Prof. A. H. Sturtevant, advisor for my Ph. D. thesis in 1942, wrote to me in England suggesting that I apply for a Guggenheim fellowship. Apparently the Guggenheim Foundation, long known as a benefactor of promising artists, writers and scientists, was lowering its sights in a moment of patriotic fervor and offering postwar fellowships to those in the services, waiving the usual necessary qualification of some demonstrated competence.

In thinking about a suitable application, I came upon the idea of testing for the possible genetic effects of microwave radiation. This was an area about which I had some knowledge and it seemed like a project worth doing. (The fact that the details of radar technology was at the time considered highly secret by the US government was a problem that I did not consider serious.) As an alternative project, I committed myself to gathering the miscellaneous data that I had collected during my graduate career and publishing them, although I had serious doubts about the advisability of doing this.

Sturtevant suggested that I undertake the fellowship (if awarded) with Prof. Curt Stern at the University of Rochester. In due course, I was informed that I had been awarded this low grade Guggenheim Fellowship. When I arrived at Rochester shortly after the end of the war I was elated to learn, by way of a graduate student in the Biology Department whose husband was a physicist, that there existed a radar set of precisely the type that I knew intimately and which put out the kind of radiation that I had specified in my Guggenheim application.

Surely Providence was smiling upon me, I thought. How could anyone be so lucky as to have chosen for postdoctoral work the one (and probably only) institution where there would also happen to be, by chance, the super secret radar set I needed? The equipment was stored in the basement of the Physics Building and was controlled by Prof. DuBridge, who had actually been in charge of radar development at MIT during the war. With great expectations I made an appointment to see him. He was a most congenial and affable individual and gave every indication that he would be helpful, but I was unsuccessful in getting any kind of commitment as to how I might go about setting up the equipment to carry out the experiments I had in mind. (For one thing, I would need a source of 24 volts, the voltage used in the B-17 bombers that I had worked on in England.) After several weeks went by I made an appointment to see him again. Once again he struck me as quite a nice guy, but was noncommittal about any actual experimental set-up. I tried for the last time several months later and was once again rebuffed. Then I realized that I was getting what might be colloquially described as "the run-around."

There might be any one of a number of reasons why I was not able to make any progress. He might, for instance, have thought that I as a biologist was basically incompetent (an attitude which many physicists have of biologists, usually for the reason that the different disciplines emphasize, even require, different kinds of intelligences). Or he might have thought that the project was doomed to failure and that he didn't want to be bothered. But another explanation finally dawned on me as probably the correct one.

Could it be that he was sitting on a hot radar set?

At our bomber base in England, the radar shack was kept under strict security. The enlisted men in my section were exempted from KP and other chores because each night they took turns standing guard over the equipment with rifles in hand (but of course with no ammunition; otherwise, someone might get hurt). Even earlier at MIT, as I was exposed to the principles of microwave radar, we had to pass by an armed guard (also with no ammunition, I'll bet) and present our special identification badges to get in.

How then was it that there was a piece of this highly classified equipment sitting in the basement of a building at Rochester, NY immediately after the war? The only conclusion, it seemed to me, was that it had been pinched from MIT. As head of the lab, he undoubtedly was recognized as the Big Boss and if a truck backed up in the middle of the night to move some of his possessions out, he would never have been challenged. It is possible, even likely, that he wanted the equipment in his possession at Rochester to continue some of the work that he had been doing at MIT.

In other words, I like to think that the radar set at Rochester probably was hot. If so, the last thing DuBridge wanted was that it be advertised that he had a radar set in his basement. So when I appeared and suggested that I work on it, he would have been dismayed, for, regardless of how the experiments turned out, there would naturally appear eventually in some journal a brief account of the experiment, along with a footnote thanking Dr. Lee DuBridge for the use of his equipment. This public announcement could cause him some embarrassment.

Sometimes I wonder—where is that set now?

Of course, an apologist for that illustrious scientist would find other excuses for his lack of cooperation. He knew, for instance, something that I did not—that during the war H. J. Muller, who had first demonstrated conclusively the damaging effect of X-rays on biological material, had done precisely the experiment I planned, with negative results. Such work would have been classified "SECRET" and DuBridge would not have been in a position to relay this information to me.

Actually, if I had been able to do these tests, all I would have accomplished would have been to cook the Drosophila. Would I have been astute enough to realize that if I could cook Drosophila I could also cook steak? Probably not. But looking over my shoulder was Alexander Fabergé, who might very well have realized this possibility.

In fact I had been presented with the basic information earlier in England and had not recognized it. One of our radar sets installed in a new bomber that had just arrived from the states was misbehaving. It worked fine on the ground but blanked out when the plane was in the air. Since we could not do any maintenance tests while the plane was in the air, we took the equipment apart, bit by bit. We discovered that in the wave guide, a rectangular pipe that led the waves from the set to the dish, which propagated the waves, there was a gasket that had been improperly positioned in the US during assembly and was acting as a barrier to the waves. The energy absorbed by this gasket had turned it into carbon and, at the change in pressure of higher altitudes, was now shorting out all the output of the radar set, preventing it from getting to the dish (the antenna) for propagation. I should have asked myself: If the waves can do this to a gasket, can it do the same to a piece of meat? I did not and so lost any possibility of inventing the microwave oven.

According to legend, a worker for Raytheon, the corporation that contracted for the radar work at MIT, noticed that keys in his pocket (some accounts say it was a chocolate bar) got warm when he was working near functioning microwave equipment. Raytheon then patented the system for home cooking. This was the same year that I was planning my Drosophila experiments, so even if I had been astute enough to put two and two together, I would have been simply too late by a few months!

A Tourist Takes in Rome

When outside the United States, Fulbright Fellows are expected to visit various labs within striking distance of their home base, discuss science, give talks, and otherwise spread knowledge and good cheer throughout the universe. Such trips are preceded by an invitation from a department or school, which, having become aware of the possibility of acquiring an outside speaker to add variety to their seminar program, sends a letter in which the recipient's contributions to science and humanity are suitably acknowledged, along with a request that he or she pay a visit.

For the children these trips can be an education. Our trips included a visit to Paris. Immediately upon entering the Louvre we were greeted by a receptionist who waved his arm to the left, apparently having been chosen for the job by his telepathic abilities, which made it unnecessary for us to ask him where we might find the Mona Lisa. Later, when Esther and the kids went to the top of the Eiffel Tower, I judiciously stayed at the bottom where I found the postcards more interesting than any sky view of the Parisian landscape. Besides, I figured that if I heard any noises indicating that the structure was crumbling (which from the historical point of view is inevitable) I could easily sprint to safety. (Among my several mental peculiarities is one that other people call acrophobia, fear of heights. I call it common sense.)

In Rome we were given the royal treatment by the head of the laboratory, whom we shall refer to only as the Professor. He took us around to the great ruins of the earlier civilizations and I was astonished at the informality and ease with which we were allowed to wander amongst the rubble. At the Colosseum I could not help thinking that if I had had a truck handy, I could easily have made off with enough boulders for several rock gardens.

The Professor took us to a little restaurant near a fountain where people traditionally divested themselves of spare change. There were no movies being made at the fountain at the time of our visit. Shortly after we settled down comfortably in the restaurant a couple entered and seated themselves not too far away. I recognized the male as a movie star I had seen many times, greatly enlarged. The woman was much younger, remarkably attractive; I figured that she must have been his granddaughter. I wondered if I could walk over, introduce myself, and ask for his autograph. I already had a nice collection of autographs of famous people, but up to now, they were all in science, and the question was whether I should expand my collection to include the arts. Anyway, they were engaged in intense conversation, oblivious to the world around them, concerned no doubt with intimate and personal family problems, and my sensitivities won out over my curiosity. I made a mental note to find out the name of the picture currently being filmed in Rome to see if his companion (presumably his granddaughter) was also playing a role, with the big screen giving me a better look at her physical endowments. Unfortunately the pressure of other matters caused me to forget this task and I never got around to it.

The next afternoon I was scheduled to give my obligatory talk, which, as I recall, was on the variations in the sex ratio at birth in humans. As I entered the lecture room, a classroom with five rows of eight seats in each, I was surprised to see that every seat was occupied. This talk was to be in English, of course, and how could it be that there were in all of Rome exactly forty persons, no more no less, who would be interested in the subject of my talk—and fluent in English too?

As I got into my subject, it became apparent that many, probably most, of the audience were in fact not interested in what I had to say. Further, I concluded by the lack of any response to my humor that they didn't even understand English. Some even seemed to have pained expressions on their faces.

What to do? Then I got an inspiration. I looked at the Professor and remarked: "At this point in my talks I like to have a cup of tea." He was obviously taken aback by this unusual interruption, said something to the group in Italian, and he and an assistant went off to find a flask and a Bunsen burner to heat up some water for my cup of tea. In the meantime the audience began milling around and slowly started to dwindle. By the time I finished the tea, which I sipped leisurely, the group had diminished to about eight or ten persons. I started again with my talk, and a good time was had by all, I think.

Anyway, I am quite sure that more than one person from that original group of forty, after escaping and returning to their more important duties in the kitchen, or in the janitorial services, eventually returned home and told their families and friends about the enforced attendance at a stupid talk in English, about how the speaker, a crazy American, insisted on having some tea in the middle of the talk, and how they cleverly took advantage of the interruption to escape and get back to their work!

Rescuing Uncle Will From His Kidnappers

It was a cold and stormy night . . . Actually, it was worse. It was one of the late autumn nights, foggy and damp, always found oppressive and unpleasant by newcomers to Oregon. We had been in Oregon only a few months and had not yet had time to cultivate that soothing growth of indigenous mold between the toes that identifies a native of the state. Here we were, Esther and I and, having been unable to get a babysitter, four kids in the car, huddled in blankets, inching our way along a rural road. We were entirely dependent on the car ahead, which we could barely make out through the dense fog, on our way to rescue Uncle Will. Two cars with the Sheriff and some deputies, all armed with shotguns (or rifles) preceded us and another car, with an unspecified driver, was behind.

It has been said that each person will have fifteen minutes of fame in his lifetime. I have heard that so many times that it must be true. I would like to add a corollary: that each person will experience at least once in his/her lifetime an event so improbable, so bizarre, that it will not be accepted by any fair-minded person as an actual happening. This is the story of our experience.

As a graduate student and friend of the Rudkin family I had become acquainted with Kathleen and Will Johnson, who were distant relatives of the Rudkins. They lived in Glendale and the Rudkins would occasionally drive over from San Marino to spend a Sunday afternoon at their house. Will and I hit it off pretty well because we had one interest in common: making radios from a miscellaneous assortment of tubes, condensers, resistors and coils.

Kathleen was a husky robust type whereas Will was frail, underweight, and clearly one who might pass away on a moment's notice. No one would predict that Kathleen would go first.

Kathleen and Will were moderately well off, having made some lucky investments in Australian mines and, having no children to tie them down, did an unusual amount of traveling. In the middle fifties, while vacationing in Hawaii, Kathleen dropped dead. Esther's father, Charles, heard about it weeks later, and when he tried to get in touch with Will, he was unsuccessful.

A couple of years later, Esther got a phone call from her father, who wanted to know if there was a road called "Lower Bottom Loop" in Eugene. After consulting a map, she assured him that indeed there was. It seems that Charles had gotten a postcard from Will. It said:

"Charlie: I am being held prisoner against my will. Can you come and get me. My address is 555 Lower Bottom Loop." The postcard was signed Will.

He gave no city or state, the postmark was smudged and all that was legible were the first two letters EU. Charles had immediately called the city of Eureka, California, but they had no street or road of that name. Then it occurred to him that it the letters EU might refer to Eugene, Oregon. When he called Esther, he hit the jackpot.

Esther, after the crisis was explained to her, called me at the lab. At first I was not too sure what to do. I called a lawyer who had helped us in another matter; he called the sheriff's office, and we arranged to try to free Will that night.

So it was that we bundled up the four kids, rendezvoused at the sheriff's office and took off for the rescue. They of course led the way, and after what seemed an interminable trip through the fog, with one false turn by the lead car, we arrived at #555. I detached myself from the rest of the family, left a safe distance behind, and went up the steps to the door. I was preceded by two sheriff's deputies with shotguns or rifles who positioned themselves on either side of the door. I knocked on the door and waited, almost quivering with fright.

The door was opened by a pleasant-looking, middle-aged woman who looked at us curiously and then invited us in. A sheriff's deputy went in first, looking cautiously around, and then stood to one side. Sure enough, there was uncle Will, sitting at a table, having a cup of tea. Within five minutes the situation was clarified. Will was clearly suffering from that mental deterioration that affects so many elderly people, and the residents were distant relatives who happened to be attending Kathleen's rites in Hawaii and took Will under their care. The sheriff's entourage slowly made their way out of the house, disappointed no doubt that their trip had not been more dynamic. Esther and I refused a cup of tea, stayed a little while longer, and then made our way home.

We visited Will a few more times before he died a year or so later. His caretakers were peppermint farmers. From them I learned several useful facts: one, that Oregon leads all the states in peppermint production and two, that goats cannot stand the taste of the peppermint plant and so are very effective at weed control in the peppermint fields.

Why I Could Never Be an Artist

Every Sunday morning while I was growing up, we got the weekly edition of the New York Hearst newspaper. One day it had in its special feature section an article on a test for color blindness that had just been concocted by a Japanese named Ishihara. It was supposed to be exceptional in detecting those who had any one of the various forms of so-called color blindness.

As I pored over the colored charts, I became aware that something was wrong. I was coming out with the wrong answers for a normal person and the right answers for one who was "red-green color blind," or, more accurately, red-green color deficient. Printing newspapers in color at that time (about 1934) was in a very primitive state; so I called to my brother Frank, who happened to be nearby reading the funnies (which back in those days were really funny), and asked him to verify my conclusion that the charts suffered from bad color reproduction. After a short time, he agreed that indeed the colors were off because he too was getting the results of a color-blind person!

In later years I would tell this tale to students taking an elementary course in human genetics. It was always followed by loud guffaws and some tittering, for the students knew (at least those who had not slept during the previous lecture knew) that if a boy had this particular characteristic the chance that his brother would be similarly affected would be 50 per cent. Thus, the fact that the two of us agreed was not so much evidence that the color printing of the newspaper was bad—just the reverse, it was strong evidence that the color was quite good!

During my second and last year at Purdue I took a physics course that included a section on the nature of light. At one point the instructor projected some colored views of spectra, some of which had been produced by shining light through cut glass vases or dishes. The girls in class seemed to be overwhelmed by the spectacular visual effects, greeting each new view with oohs and aahs; I on the other hand was completely unimpressed, being barely able to make out any colors at all. This puzzled me; perhaps the instructor was playing some kind of joke on the class, so I went up to the front after class and asked him to show the slides again. Up close I could see that indeed there were colors there on the screen, although certainly not the sort of thing I would write home about (actually I practically never wrote home about anything!). The instructor suggested that I visit the physiology professor and have him test me with Ishihara charts again. This time there was no equivocation—I was indeed red-green deficient.

Almost Failing Chemistry

During the preceding year as a freshman at Purdue, I had taken an elementary course in chemistry for chemical engineers. It consisted entirely of calculations during which we had to tackle problems like: if a crop requires 12 kilograms of nitrogen sulfate per hectare each year for a good crop, how many tons of manure would be needed each year on ten acres of farmland that is only 85 per cent efficient? The instructor (Nelson, I believe his name was) insisted that all this be done with the hands never leaving the slide rule. Resort to pencil and paper would have been grounds for instant dismissal.

I remember him distinctly because he was well known for his unwillingness to assume responsibility for female graduate students. His argument was that chemical syntheses require long hours of arduous watching of reactions as they boil and sputter, and that it was dangerous to give this task to women, who would have to leave their work to visit the women's room because their anatomy was such that they could not pee in the sink, as the males regularly did!

The calculations I could handle, but what did me in was the chemistry lab. Here we were asked each week to go through a series of synthetic steps and we had to determine the acidity of our preparations using litmus paper. As I recall, the system was something like this. You dipped the paper in the solution and if it turned bluish green or light pink it was acidic. If on the other hand it turned greenish blue or dark pink it was alkaline (I think). Needless to say I was completely befuddled. Mostly I got my answers from concerned lab partners who kindly told me what the colors were. This was truly altruistic on their parts since this was a class made up mostly of highly competitive pre-med students fighting for good grades to improve their chance of admission to medical school. Nevertheless, despite their helpfulness, I flunked (or at least did very poorly on) the lab section of the course. This experience was a disaster and I remained at arm's length from chemistry for the rest of my life.

I was greatly heartened when, upon coming to the University of Oregon I learned that one chemistry professor wisely gave his students color blindness tests before subjecting them to the ordeal of using litmus paper. Of course, these days, instruments are available to do the same job even better but I was born too early for that.

The red-green color deficiency never really causes me great trouble. I can distinguish red from green traffic lights without any problem, unless they are quite distant and faint. Anyway, the red light is at the top and the green at the bottom (except in Wheeling, West Virginia in the 1960s!) Curiously, my greatest problem is with the blues and greens.

One autumn evening I was sitting at the dinner table when my younger daughter Ellen asked:

"Daddy, where's our car?"

A foolish question, I thought; I could see the car clearly out the dining room window.

"Out in the driveway, can't you see it?"

I replied, patiently, as a parent must to a child's stupid remark. "No," she said, "Our car is blue and that car is green."

We went outside to inspect, and she was right. Apparently I had driven home from the university in the wrong car!

So I drove back and as I approached the spot where I had parked I could see not only my blue car, but an obviously distressed young male pacing back and forth. When he noticed our arrival he dashed up and said "What the hell are you doing in my car?" I tried to explain that because of my color blindness I had mistaken his car for mine but he was not impressed, still suspecting some kind of skullduggery, still belligerent. I then insisted that he take his key and try it in the door of my car, an almost identical model, which was sitting in the next space. His key worked. Then he tried it in the ignition of my car and the motor turned over without hesitation. Still not happy, he got into his green car and drove away.

This is the closest I ever came to being beaten up because of my red-green color blindness.

SOCBAG

Sturtevant and I shared at least one trait in common: we were both red-green color deficient. He formed the Society Of Color Blind American Geneticists (to which he applied the acronym SOCBAG). Besides the two of us, his membership list included Max Delbruck, E. H. Grell, L. H. Snyder, D. R. Stadler, and Dwight Miller, with R. B. Goldschmidt and Felix Bernstein as Posthumous Foreign Members. He proposed to send out issue # 1 of the Color Blindness Information Service (CBIS), which would take up the (then) recently reported association of color blindness with alcoholism. Several years later I received the second (and last) issue which is here presented in its entirety: "Heavy doses of alcohol lead to temporary color blindness. If alcoholics are tested when they've been off the stuff for a while, they show normal CB frequencies. So we can breathe easily again. AHS."

A Most Unusual Character

Everybody knows somebody whose characteristics are well beyond the normal curve. For me, that spot is taken by Alexander Cyril Fabergé.

I first met Fabergé in London on 24 July 1945. The war in Europe had ended and we were waiting to be reassigned to units destined for Japan in preparation for the bloodbath that was predicted for the conclusion of the Pacific war. The incessant bombing of London by the German's V1 and V2 weapons had made the place so dangerous that American personnel were not allowed to go there on leave during the hostilities, and so when I read that the Genetical Society, an English counterpart to our American Genetics Society, was meeting in London, I looked forward to getting back into genetics for a bit. Also, a trip to London seemed an appropriate way to celebrate my birthday.

I think that I was the only American at that meeting, but the place was overrun with British geneticists who, I suppose like me, were celebrating their first freedoms from the restraints imposed by the excesses of the war. During this meeting I met some of those English scientists with whose work I had become familiar during my graduate work for the Ph. D, among them M. J. D. White, Cyril Darlington, and J. B. S. Haldane. I did not meet R. A. Fisher, but heard his lecture on the complications of genetic segregation in plants with more than three sets of chromosomes, a subject whose complexity was matched only by its dullness.

I do not recall how Fabergé and I got together at that time; perhaps it was arranged by someone who knew that both of us had been involved with radar in our respective services. Anyway, we got along fine and instituted a friendship that lasted until his death a few years ago. An insight into his life and intellect may be found in the article entered under his name in the Bibliography.

Fabergé was the grandson of the Egg Man of the Tsar. At the time of the Russian Revolution, when he was four or five, he escaped with his family into Finland, then to Switzerland, Paris, and England. Finally, he made it to the U. S. Our paths crossed continually. He spent time at the Universities of Wisconsin, Missouri, and Texas; we were together at Missouri and at the Oak Ridge National Laboratory. At Missouri we collaborated on a course in Advanced Genetics, he doing the plant end of things and I the animal. I learned much more from sitting in on his lectures than I ever did in the graduate courses I took at Caltech with Sturtevant and Dobzhansky.

He was a perfectionist in all that he did, particularly at the microscopic level. For instance, when I was doing some experiments trying to verify the results of others that the X-ray mutation rate increased markedly when the specimens were at a very low temperature, a result that seemed contrary to common sense, he helped me determine the actual temperature of flies exposed to below freezing temperatures by making a thermocouple (a temperature-measuring device) so small that I could insert it up the rear end of D. melanogaster (a very small fly) without injury.

One of his notable contributions to Drosophila technology was the design of a food pump, fashioned to squirt into food vials precisely the amount necessary for proper culturing, or alternatively, four squirts per quarter pint culture bottle. Dan Lindsley tells me that one of the original prototypes is still in use in the Drosophila laboratory at the University of California at San Diego.

He had an extraordinary intellect and a memory beyond compare. He was without family, and could spend enormous amounts of time in the library. He was knowledgeable about the latest achievements in physics, chemistry, astronomy and other sciences as well as those in biology, and would be asked for technical information by workers in those other areas.

Fabergé never seemed to settle down at one place. Part of this might be attributed to his complete disdain, openly expressed, for those in the academic hierarchy who, being devoid of any academic merit, climbed the professional ladder by virtue of politics or of carefully nurtured personal relationships. This attitude of his did not sit well with those higher in the bureaucracy who were responsible for promotions, tenure, etc. There are however two other reasons for his lack of success, as viewed from outside. One was his preoccupation with perfection, which caused him to waste much time on inconsequential aspects of a problem, and the consequent failure to produce a substantial publication record. Administrators, like historians of science, like to measure academic research achievement by making a precise count of the number of publications a person authors.

He was, for instance, the only one I have ever known who before inserting or removing a screw, would stare at the screw head, assess the length and thickness of the slot, and then go to his array of screwdrivers and select the one that would fit exactly. He was so annoyed at the University of Texas by colleagues who would, in his absence, go into his lab to "borrow" a screwdriver that he bought a cheap set that he kept in full view for any to take.

The second was that, to put it gently, his behavior was at times so peculiar that many thought that he was as nutty as a fruitcake. Twice when I went off on sabbatical I invited him to Eugene to give the genetics courses in my absence, and this gave the graduate students a good chance to see him in action. Their first encounter in the biology building during the evening hours was apparently a bit startling. One description given me later went something like this: "As I walked down the hall towards his office, I could hear several people arguing violently, and as I got closer, I realized that two people were involved, one of the two was speaking in French and the other in Russian. But when I passed by the open door I could see that only Fabergé was in the room."

When he was working on a construction project and things were not going well, he would address the tool in a very personal and somewhat unfriendly way, usually in Russian. And, as he was walking along the street, he had no difficulty in carrying on an animated conversation with himself, a habit that caused a certain amount of surprise and amusement on the part of passing undergraduates.

But he was intensely interested in all kinds of problems. At one point he brought up the question as to what happened in plants that are self-sterile, but have three distinct forms, each fertile with the other two. In its simplest form where each of the three types has a different set of alleles (like AA, Aa and aa) the mathematical answer is so simple as to be trivial. But for a different simple case where there are two sets of genes involved (A and a alleles on one chromosome and B and b alleles on another) the answer gets sticky. He enlisted the help of D. J. Finney, a very competent mathematician and statistician he knew from England, and me (not so competent). Finney published several papers on the general problem. I got hung up on the complications of the mathematical formulation. I could, by computer simulation, show that the three types would, after several generations of random breeding, settle down at a frequency of one third of each of the three types. But even knowing the answer, I could not show that the basic equations were consistent with that answer. Finney suggested that a mathematician could see, by inspection, that the result would involve an equation of the 27th degree, whereas my intuition (for what that is worth) told me, that the solution must be simpler, since the numerical coefficients in the original equations were all multiples of one-fourth and the unknowns always occurred singly or in pairs.

This problem was a challenge, a welcome diversion from the routine of other work, and it has occupied my attention, on and off, for several decades. After tedious algebraic manipulation I was able to reduce the system from five equations and five unknowns to an equation involving a single unknown in the fifth degree, which, as is common knowledge, is insoluble for the general case. I tried in vain to bring the equation down to a lower degree by finding some factor by which my equation was divisible. Finally, after reading that some mathematicians at the University of London in England had developed a computer program to seek out the roots of polynomials, I sent them my final fifth degree equation and they concurred that it was unfactorable.

I am still puzzled by the apparent complexity of what should be a simple formulation, and will get back to the problem when I have more time on my hands. I still have in mind a few new approaches that are now possible with the development of powerful computer programs that can almost do the impossible. Thanks to Alexander Fabergé the attempt to simplify this problem has become something of an obsession with me, much like the fruitless attempts of the ancients to square the circle, or to find a simple rational value for pi.

Celsius and Linnaeus

As I entered the foyer of the building at Uppsala University I was taken by surprise at the lavish floral welcome. The Swedes are well known for their friendliness and hospitality, but this welcoming was a bit much.

I had come to this school to visit with the Historian of Science in order to learn more about the sequence of events involved in the change of nomenclature in the measurement of temperature, and in particular was interested in finding out (something which I discovered later every Swedish schoolboy knows) why they the Centigrade scale was named for the wrong person—the one who had the scale upside down—and not for the one who had it right.

At one time there was quite an effort to convert the US (or English) system of weights and measures into the metric system. It was common to see along the roadside signs giving the distance to the nearest town in kilometers, or in both miles and kilometers. This conversion was to include the change in temperature readings from Fahrenheit to Centigrade, or Celsius. This latter move irked me somewhat. An argument could easily be made for the change to metric lengths and weights. The increasing commercialization, globalization and homogenization of our every day lives (and therefore of us too) would be accelerated if all people throughout the globe would live by exactly the same rules.

The proposed change in thermometer scales seemed to me to be foolish. We would be exchanging a useful scale of long standing familiarity for one unfamiliar and much cruder, because the degree on the Centigrade scale is almost twice as large as the one on the Fahrenheit scale.

So I decided to look into the matter of the reasons for the change, and what I learned surprised me. The trip to Uppsala was part of this effort to understand the sequence of events.

To begin, I knew that the metric system had its beginning at the time of the French Revolution. It seems that the French scientists of the day were fearful of the guillotine and as a group decided to show the revolutionaries that they were in fact useful citizens. So they decided that all measurements should be based in the number ten. The meter length was supposed to be one ten millionth of the distance from the equator to the poles. The measure of weight, the gram, was the weight of one cubic centimeter of water and successive multipliers of ten would give the greater weights.

These have become part of the system of weights and measures; other suggestions have been treated less favorably. Time was to be divided up with a base of ten instead of twelve, and a quarter of the circle was to be divided into 100 centesimals instead of 90 degrees. These two proposals mercifully died natural deaths. I do recall, however, that during my first year at Purdue I was required to enroll in ROTC. We were an artillery unit and the only piece of armament available was a French gun of WW1 vintage. We learned to point it in the right direction (hopefully) not by using a scale in degrees, but one in centesimals. My military career at Purdue was cut short when I became allergic to the wool used in the government issued uniform. In retrospect, I wonder if that sensitivity was more psychological than physical.

Back to the temperature scales. Both Carl Linnaeus and André Celsius were faculty members at Uppsala University during the first half of the eighteenth century. The first is universally known for his work on the classification of plants and animals, the second was a distinguished astronomer and meteorologist. Although at opposite ends of the political spectrum (the biologist being liberal and the physical scientist conservative), they were fast friends. Both used thermometers based on one hundred degrees between the freezing and boiling points of water. But Celsius' thermometer had the boiling point of water at zero degrees and freezing at one hundred whereas

Linnaeus's had it right, just the reverse. Linnaeus published an account of his thermometer more than two years before Celsius did his. Why then was name of Celsius attached to the thermometer scale rather than that of Linnaeus?

It seems that during the 1960s the World Meteorological Organization considered the report of a committee on nomenclature that suggested that the centigrade thermometer be renamed "centesimal." Why is not clear. But then it was pointed out that the word centesimal had already been preempted as a substitute for the degree as a measure of an angle and could not be used. One person, clearly a Swede, volunteered a way out of the impasse. Why not, he said, name it for the great Swedish scientist Celsius? Everybody heaved a sigh of relief and the suggestion was passed unanimously.

The fact that Celsius used the scale upside down was irrelevant. He was in fact a most accomplished astronomer and his duties included weather observations including temperature, throughout the year. And at that meeting in the library of Uppsala University I got the answer to my question: why did Celsius have his scale upside down? The answer proved to be very simple. Celsius, as part of his duties, kept track of the daily temperatures at Uppsala. In the winter these often hovered around freezing. This would require many entries of plus and minus one two or three degrees, and the probability of errors becomes significant. By using freezing at 100 degrees this problem vanishes, since a one degree variation from freezing would be either 99 or 101 degrees rather than +1 degree or -1 degree.

And I also learned the reason for the magnificent floral display in the lobby of the Library at Uppsala University. Just by chance, I had walked into that building on the anniversary of Celsius' birthday, and the flowers were meant for him and not for me!

SECTION 2

THE EARLY YEARS

Into the Eye of the Storm

I was led by Theodosius Dobzhansky down a set of steps into the basement of the Biology Building, Kerckhoff Lab at Caltech, and was about to be introduced to one of the most eminent geneticists of the day, A.H. Sturtevant.

It was pretty hot that September day in 1938; the coolness of the basement did little to ease my sweaty apprehension. Even as a high school student, and later at Purdue University, I had read all of his scientific works I could get my hands on, and, having the "Dutch book" (Morgan, Bridges, and Sturtevant 1925) as my main source of information, considered him to be one of those three truly great contributors to the birth of the science of genetics in the United States. Naturally I felt uneasy about what effect this meeting might have on my future career as a candidate for a Ph.D. in genetics. I certainly was not prepared for the somewhat cool reception I was about to receive.

We entered a dimly lit room, at one end of which was a desk lamp illuminating a microscope and before it a dim figure. He looked up as we approached, peering over his glasses at Dobzhansky and me. The introduction was short and monosyllabic, even curt, after which Dobzhansky proceeded to tell Sturtevant of the plans for a research project that he had decided I should pursue in fulfilling the requirements for the degree. Sturtevant remained silent during this exposition. Every twenty or thirty seconds he would reach into a pail on the floor beside him, pull out a small piece of broken ice, and hurl it against the concrete wall some twenty feet away whereupon it would smash into smithereens.

This performance with ice fragments would have seemed peculiar to most, but to anyone familiar with Drosophila it was completely understandable. Sturtevant was simply trying to provide primitive air conditioning during the heat of a miserable September day in Southern California, not so much for personal comfort as for the preservation of his many fruit fly species lined up in dozens of culture bottles on shelves against the wall. They would have died or at least become sterile at higher room temperatures.

At the conclusion of Dobzhansky's explanation, he simply said "Well, I guess that's all right," and we left, taking that as his stamp of approval as well as his dismissal of us. At no time did he express any interest in me or any personal problems I might have as a newly-arrived first year graduate student. I was clearly Dobzhansky's responsibility and that was that.

Although I was puzzled by the meeting, I innocently attributed the apparent lack of interest on Sturtevant's part to his preoccupation with some thoughts that we had interrupted. I also reasoned that a scientist of his stature was entitled to some idiosyncrasies. A person more perceptive than I might have wondered whether he was hurling the ice against the wall with more vigor than was necessary. It was not until years later that I realized that Sturtevant might have had, on the one hand, a strong reason for feeling a personal curiosity (see Endnote (1-1)) in my arrival and, on the other, a feeling of annoyance at my appearance on the scene.

What I could not have known then was that I had blundered into a complicated situation with undertones that nearly meant disaster in my efforts to get a graduate degree, and, more importantly, eventually led to Dobzhansky's departure from Caltech. These times would prove to be so traumatic to Dobzhansky that, years later, when he set down his memoirs (Dobzhansky 1962) and recalled in exquisite detail the events of his childhood, of his first years in America and of his later life, he left completely blank a period of more than four years that included the time immediately preceding his departure from Caltech. This omission is handled with such finesse that even the most astute historian of science, upon reading his memoirs, would not be inclined to wonder why these years were not discussed, or even, for that matter, to notice that they were missing.

This discussion is continued in Section 5 to which the reader should jump if his primary interest centers on the relationship of Sturtevant and Dobzhansky. The intervening chapters describe the author's early experiences, provide a background to understand how his particular allotment of the various intelligences shaped his scientific career, and show how these put him into association with both of these men and their works.

Recollection of My Grade School Days

Some writers can recall not only their first years of childhood, but even their experiences in utero, including the cataclysmic impact of the sperm on the egg. My memory banks are not that retentive.

I do recall in my early years constructing a cart from an old abandoned baby carriage and riding it down a sloping sidewalk in front of the grade school near our house, not anticipating the consequences of meeting, head on, the concrete wall at the bottom.

I also remember the very first day of school when several of us, overwhelmed by the excitement of finally getting into the first grade, failed to stop chattering after being asked several times by the teacher, and then were subjected to several sharp cracks of a ruler on our palms.

During those grade school years, I cannot say that I was a happy child. In particular, if, as I was putting on my shoes in the morning I happened to think of what a stupid business living was, I would fall into a deep funk from which I would not recover for several hours. Fortunately, about half the time I would not, during this critical shoe tying period, think of the inanity of life and so would have a reasonably pleasant day.

I remember walking along a cliff skirting the Susquehanna river and discussing with a friend (name long forgotten) the flight that Lindbergh had just successfully completed over the Atlantic and sagely predicting that in a few years such flights would be commonplace.

I remember socking my best friend in the nose when he exulted in the election of Herbert Hoover over Al Smith. He did not know that I was a secret advocate of Smith. This friend was James Fritz, one of those child prodigies who had already skipped several grades in elementary school because of his precocious abilities.

About the time of entering high school, I had some juvenile experiences with science. I enjoyed puttering around with electricity and radios. I spent hours making hydrogen from sulfuric acid and zinc, and hydrogen and oxygen by the electrolysis of water. It was here that I got the great idea of deriving a great deal of energy from water, since hydrogen mixed with oxygen can react with explosive force to form water. This was such an important concept to me that I wondered whether I should apply for a patent, but then I realized that an idea so important would surely be stolen by the big companies, and the best thing to do was to keep it a secret until I could get a working model going (which, of course, I never did).

These were the early days of radio; I accumulated a multitude of parts, condensers, resistors, vacuum tubes, etc. and constructed many combinations of components. I remember in particular making a super-regenerative radio receiver, which, when the coils were too closely coupled, would emit an unearthly howl, heard no doubt by many radio listeners for blocks around, since my receiver (consisting of one 201-A vacuum tube) was now acting as a transmitter. My crowning achievement with this radio was the reception (in Wilkes-Barre, Pennsylvania) of a station operating out of Mexico (with studios in Del Rio, Texas) in which a Dr. Brinkley regularly urged men whose sex lives were faltering to visit him for an operation which would replace their defective sex glands with those from (if I am not mistaken) a goat.

During those early years I enjoyed making maps, redrawing some I got from books to different scale. The physical similarity between the outward bulge of South America and the indentation on the western coast of Africa looked to me as though the two had at one time been joined together. I thought this to be an original idea, unaware that this had been proposed many years earlier in great detail by others.

At one point there was, of course, the inevitable butterfly collection. From Holland's butterfly book I learned the scientific names of some of the most common that I collected—Papilio (the swallowtail), Vanessa antiopa (mourning cloak), Colias (the small white cabbage butterfly), and a few others. These names stuck in my memory so that in later life when someone questioned my qualifications as a biologist (which they had good reason to do), I would casually reel off a few of these names at an opportune time and thereby reestablish my credentials. At the same time it gave me a passing acquaintance with that aspect of biology that dealt with naming things, and led me to dread any exposure to disciplines such as taxonomy, etc. where this was the primary emphasis. In fact, if there was any force in my childhood that worked against my adopting biology as a career, it was the horrible thought that I might have to memorize scores of scientific names. At the same time I realized that some aspects of biology that were highly descriptive, embryology, for instance, had no appeal to me whatsoever.

Over a period of time I had in my menagerie an owl, several rattlesnakes, and an opossum. The latter gave rise to an unusual out-of-earth situation. I caught the opossum in early spring and kept it in a cage in our basement. It seemed unusually docile, even allowing itself to be petted on the head, provided that the hand approached its head from behind. It was happy (I thought) until one day the cage was empty. It had escaped but it was nowhere to be found until some weeks later my mother, upon opening a built-in cabinet door, found it sitting peacefully on the top shelf. It apparently had been living in the partitions between the walls and had found a loose board into that cabinet. It immediately scurried back into the wall and we saw it no more. However, a short time later my father's job changed and we moved to another part of town. Somehow word got back to me that the family that moved into our old house were able to stay in it for only a few weeks, abandoning it because of supernatural events. They claimed that they heard strange noises coming from the walls of the house and that surely the house was haunted. It dawned on me that this was springtime, the time when the young of most wild animals are born, and that I may have had in my cage a pregnant female opossum who decided to nest in the walls of the house instead of heading out for the woods. However, this was a very iffy hypothesis, and I preferred the ghostly explanation. Besides, I had read that the opossum, as a marsupial, kept its young secure in her pouch, and so, I reasoned, the young would not necessarily have to be born in the springtime.

How Not Going to Church Got Me into Biology

Of all the miserable experiences of my childhood, none could compare with those connected with the church. We lived in a predominantly Catholic, largely Polish community populated mostly by poverty-stricken families in which the breadwinner, working in the anthracite mines, was engaged in an unsuccessful fight to maintain some semblance of decent living for his family. Going to church was a necessity.

There were two main problems with this. First was the matter of the confessional. This was an art that I never did master. During those pre-adolescent years I had great difficulty rounding up enough sins to make a confession worthwhile, and it was necessary for me to fabricate a number of nonexistent ones to justify a trip to the confessional box.

On the other hand, there were on occasion some events that I considered so horrendous that I would keep them buried in my guilty little mind and would not mention them to anyone, particularly not to a priest who would surely be unsympathetic with my mental torture.

There was, for instance, the time that I was chasing a classmate (for some reason long forgotten) who, in trying to escape, ran out into the street, was struck by a car, and spent several weeks hospitalized with a broken leg. Another time a pal of mine and I were throwing rocks at each other from a safe distance. I threw one to a spectacular height and when he saw it coming, he ran as fast as he could. Unfortunately his perceptions of distance and direction were deficient and the trajectories of him and the rock intersected. He went, unconscious, to the doctor's office.

Sins such as these were not meant for public discussion but were to be held within that place in the cerebrum marked "guilty remembrances." Certainly they would not be discussed with a stranger like a priest at a confessional. I would rather have burned in hell.

But the most unpleasant aspect of religion for me was going to church in the summertime. The religious services were in Latin and whatever was not done in that tongue was in Polish. My parents were very strict about not using the Polish language around the house, and so I was deprived of fluency, or even acquaintance, with a second language. I had not the foggiest notion of what was going on during the services. And before the days of air conditioning the heat was unbearable in the summer.

Under the circumstances it was necessary to exercise some individual initiative. After being escorted to the church by my older sisters, I would seat myself at the rear of the church and at the first opportune time, would sneak out and go for a walk, timing myself to return to my seat before the services terminated. This of course added to my store of guilt. (I do not recall ever confessing this sin to the local priest). However, I was greatly relieved when on one of these walks I ran into my brother Frank (older than I by two years), who was doing precisely the same thing.

My High School Years

Matters turned in a biological direction when my brother Frank told me that there was a superabundance of water snakes to be found along the railroad tracks that skirted the Susquehanna river. Thereafter, every Sunday morning when I skipped church, I would make my way down to the river to see what I could collect.

After the initial encounter with water snakes, I extended my collecting and would overturn every log and rock of reasonable size to see what was underneath. I became aware that in the hills surrounding the town of Wilkes-Barre there were quite a few rattlesnakes. With a classmate, Norman Nogic, I roamed the hills with a cloth bag and a forked stick and did in fact catch one on each of several trips, although our searches were usually unsuccessful. In retrospect those trips seem remarkably foolish; if one of us had been bitten the chance of making it back to civilization for help by way of a steep footpath of several miles now seems rather dismal. In our adolescent minds the piece of cord that we carried, to be used as a tourniquet, was adequate protection against any calamity. At some point we included Eddie Lewis in these trips, but usually the collecting forays were made by me and Nogic. We brought our captives into school, the Elmer L. Meyers High School, where the biology teacher Steve Emanuel provided a locked cage in a locked biology prep room behind the biology classroom.

At first, during vacation, I would bring them home, but this did not meet with the complete approval of my mother and my four sisters, so the snakes generally stayed at school. On several occasions we tried to feed the snakes by putting live mice in the cage. The rattlesnakes would kill the mice, but would not swallow them. So, at intervals of a few weeks, Mr. Emanuel would bring in some raw hamburger and I would hold the rattlesnake by the neck as he forced some of the meat down its throat. These were exciting times and the adrenaline ran high. I do not recall what eventually happened to the snakes; I do not remember ever returning them to their natural habitat. In those days it would never have occurred to us to let them go.

At the suggestion of Mr. Emanuel, who, in addition to being the biology teacher, was also the athletic coach, I became "manager" of the football team, which meant that after each game and practice session, I would collect all their dirty, sweaty uniforms and bag them for the trip to the laundry.

And the director of student activities, Miss Hogg, asked me to serve as editor of the school paper, the ElmPrint, a job that involved mostly cutting up copy and deciding where and how it would appear on the final printed form. I exercised no editorial responsibilities.

Outside of school, I worked in a small print shop. It derived a good deal of its business from printing lottery tickets, the winning numbers of which depended on the figures for the U. S. Treasury reports, which appeared on the front page of the daily newspaper. I think that this printing was, strictly speaking, illegal, although not pursued vigorously as such by the local authorities (3-1).

My interest in science, almost an obsession, manifested itself in another way. I thought that it would be an achievement to get the signatures of well-known scientists. Writing them and asking for an autograph was, of course, too crude an approach. My procedure was to look in the encyclopedia for their biography and an account of their work and to find some inconsistency, minor error, or incomprehensibility which could form the basis for a polite letter of inquiry. In this way I acquired about half a dozen signatures; their letters and names have been lost over time, although I do recall getting brief but polite replies from the British astronomer Sir Arthur Edington and the physicist Sir William Bragg.

I Almost Became Valedictorian

When I was a sophomore I became entranced, from a distance of course, by a girl named Lois Andrews who was one grade below mine and who, therefore, was in classes one level below those I was taking. I desperately wanted to get into one of her classes in order to get closer to her.

I finally got a great idea. She was taking Latin as a foreign language; I myself had chosen French. She would be taking the second year of Latin the coming year and maybe I could get into that class if I mastered first year Latin over the summer. So I got the prescribed text and during the summer I covered all the stuff for the first year. Sure enough, at the beginning of the next year I was able to get into her class!

It turned out that she was not particularly interested in me and I found out that she was not so great after all. Anyway, she was not nearly as cute as somebody I found later thanks to Th. Dobzhansky. However I had two years of Latin as a result and have been a great Latin scholar ever since, being able, during dinner conversations, to pass off phrases like *tempus fugit* and *semper fidelis* with considerable nonchalance.

I thought no more about this until near the end of my senior year when my French teacher, Miss Tyburski (3-2), told me that the school office had added up the credits each student had accumulated, which they calculated by totaling the product of the grades for each course by the credit for that course. I had a higher total number than anybody else, because of the extra courses I had taken (Geometry, Calculus and Latin), Latin had put me over the top. This meant that by their rules I would be valedictorian.

I panicked. The idea of getting up in front of all those people at graduation time and making a speech terrified me. I did not know what to do; running away from home and hiding until after graduation seemed to be the only way out (3-3).

However it turned out that some of the teachers were also upset. They knew who the really bright students were—those who always got A's (I had several B's over the years). These were the teacher's choices and they were not going to let some inferior student (namely me) take the prize because of some fluke. They hurriedly had a meeting of all the teachers the next day and revised the system of calculating the standing. Instead of basing it on the total number of points, which was stupid anyway, they calculated the average grade (GPA). So it was that everybody ended up happy!

A bunch of us were members of a chess club, once again including Lewis, Nogic and me. After several weeks of playing each other, we decided to challenge the team that was sponsored by the local YMCA. A date was set and we all showed up at the YMCA for a set of matches in which we would pit our skills against theirs. However when we arrived we discovered that somehow either we or they had gotten the date wrong, and the only person present was a lone, forlorn-looking kid, apparently their mascot, who was engaged in cleaning up. He seemed to be a good sport however and volunteered to play all of us (five or six, as I remember) simultaneously. Not wanting to have wasted our trip, we agreed. In short order it became clear that he was good, very good. In fact, he beat us all. He offered to play another set. We said no thank you and made a hasty exit. We did not schedule a rematch.

Getting into Fruit Fly Genetics

During the high school years we (Nogic, Lewis and I) spent much of our free time at the public library. It was called the

Osterhout Library, one of the many set up by Andrew Carnegie in his effort to salve his conscience by getting rid of his ill-gotten gains. It was a regular trip, always walking, and never a pleasure during the bitter cold of some of the winter months. The first order of business was to go to the journal shelf and look for any new arrivals, then to the new book shelf, and finally to the science section to see if there were any books that had previously escaped our attention. I believe that all three of us read every science book on the shelves. I remember two in particular, one by Ernst Haeckel called, I think, The Riddle of the Universe which introduced to me the phrase "ontogeny recapitulates phylogeny", and another by a president of Cornell University named White on The History of the Warfare of Science with Theology.

At Steve Emanuel's suggestion, as juniors we joined the biology club. There never were more than four or five members that year and Eddie Lewis, Norman Nogic and I were the only three constant participants. It was such a small and closed society that we decided after a while to rotate the officers' positions, each taking the post of president, secretary and treasurer in turn, dutifully switching every month or so.

At one point Lewis reported at a meeting that he had just been to the library and had seen an ad (Figure 1) for the sale of fruit flies that could be used for experiments, and he suggested that we get some. This started the sequence of events that led to Lewis's and my ending up as geneticists, and, most spectacularly, with his sharing with two others the Nobel Prize in Medicine in 1995.

As treasurer in full control of the club funds I naturally had the job of sending in the order. We had only about two or three dollars in the treasury but it was enough to order one culture. The ad, however, described a number of different varieties—white eyes, vestigial wings, etc. and so to cut down on the expense I cleverly asked that several of these types be included in one culture.

9V14 Drosophila, the Fruit Fly. The Drosophila fly is famous through its use in Genetics to demonstrate the Mendelian ratio. It is also valuable for use in the laboratory to demonstrate complete metamorphosis of a typical insect, as the entire life cycle is completed within a short time. With each shipment we send a detailed instruction leaflet telling how to prepare media, mate the flies and make crosses.

The following stocks are suggested as being most satisfactory for general work, but many other strains can be supplied from our laboratory stocks.

Sex Linked

white (eye)
eosin (eye) minia-
    ture (wing)
bar (eye)
miniature (wing)
yellow (body)
eosin (eye)

Third Chromosome

ebony (body)
sepia (eye)
spineless (leg)

Fourth Chromosome

eyeless (really a re-
    duced eye)
Wild or Red (eye)

Second Chromosome

vestigial (wing)
brown (body)
curved (wing)
lobe (eye shape)

Special Stocks

5 sex linked characters in combination.
5 second chromosome characters in combination.
5 third chromosome characters in combination.
Lobe—curly stock.
Dichaete—hairless stock.
Price, per culture with instruction leaflet $2.00

Figure 1. The ad from the Turtox Biological Supply Company as it appeared in the journal Science in 1934. It describes the availability of Drosophila stocks for simple laboratory experiments. From this one can see why there was some ambiguity about precisely what one got for the sum of two dollars, and why, with only two or three dollar in the "treasury" of the Biology Club at the Meyers High School, I ordered one culture which carried a mixture of several simple mutant characters. S. A. Rifenburgh of Purdue University, who was responsible for the ad and for supplying the fly cultures, wrote me explaining my error. He sent the additional cultures gratis. His offer of help led to an extensive correspondence, which culminated in my getting a scholarship to attend Purdue, followed by his suggestion that I correspond with C. B. Bridges, and eventually my decision to do graduate work at Caltech.

After a couple of weeks I got a reply, not from the Turtox Biological Supply Company, but from Prof Sumner A. Rifenburgh at Purdue University, who was apparently augmenting his meager professor's salary by filling these orders. In a long handwritten letter he explained that a culture was defined by only one mutant type and since I was obviously a beginner he would send me several cultures of mutants free. Further, he volunteered to give me any other advice or information we might need.

At that point we were able to carry on a few elementary crosses, although this was done with some difficulty because of our inexperience with culture methods and mold contamination. We were more or less on our own. I do not recall to what extent the others pursued this interest beyond the elementary crosses. I myself did vigorously, encouraged by Rifenburgh with whom I had a continuing and extensive correspondence on these genetic matters.

Not having any instrumentation available other than a simple hand lens, I was fortunate in spotting a mutant with wings not laid flat against the back as is normal with most flies but with the wings held out at right angles to the body, something completely obvious without any magnification. In one of those rare moments of inspiration which come to any human only several times in his lifetime, I named the mutant "held out." I would have never guessed that it was destined to become one of the most interesting mutants of Drosophila melanogaster, that this locus would eventually be honored with the more scientifically pretentious name of decapentaplegic and that a corresponding gene would eventually be found as part of the human makeup (4-1)! During the summer between my junior and senior years in high school, I was aided by my younger sister Glory, who helped as chief cook and bottle washer, and who really made this summer work possible.

The next step in this work was to find out if this mutant was a repeat of some well-known mutant or a brand new mutant and, if the latter, where in the fly's four chromosomes the responsible mutated gene was located.

My correspondence with Rifenburgh on the matter of "held out" was voluminous and, of course, came to include many personal topics as well as genetic ones. On the matter of going to college, he suggested that he could try to get a tuition scholarship for me and that I could then go to Purdue. This of course was a godsend to me because without some outside help I had no chance to get to college, and especially not to one of the stature of Purdue University. He succeeded in his efforts and in due course I was off for my great adventure. (Similarly, Chris Bayes, one of my brother Frank's teachers, succeeded in getting him a scholarship to Lehigh University.)

It was through Miss Hogg that I was introduced to a Mr. McClintock, who provided me with a scholarship from the McClintock Foundation, a modest $100, which was sufficient to pay for my bus fare to Purdue University in Indiana with enough left over to sustain me for quite a while.

My College Career

I like to try to impress my grandchildren (and others when the situation warrants) with the statement that I entered college at age 18 and got my degree at age 19.

This is, strictly speaking, true, but is misleading because it leaves the impression that I got a bachelor's degree in one year. Actually, because my birthday is in July, it took two years, and not because I was one of those child prodigies who appear proudly in the magazine section of the Sunday newspaper. It just happened that I accidentally found a loophole in Purdue's requirements for graduation. It came about in the following way:

Because I had received credit for all elementary math courses in high school, my first math course at Purdue was to be calculus, the elements of which I had mastered in high school. It was with some disappointment then that I discovered, at the first meeting of freshman class, that the text was to be "Calculus Made Easy," the very same text I had used in high school to get additional credits. When I explained my disappointment to the instructor, a Prof. Marshall, he explained that the Rules and Regulations of Purdue University provided that credit towards graduation could be given if the student passed an exam covering the content of the course. The rules went much further—they stipulated that any student applying to get credit by exam must be provided with copies of all the exams given in that course during the preceding several years. It took no trouble at all to satisfy Prof. Marshall that I had mastered all (or most) of the material in that calculus book.

Then it dawned on me that I had hit upon a gold mine, for there was no reason why I should not accumulate credits towards graduation in other subjects as well. I was strongly motivated by the feeling that a bachelor's degree was nothing more than an impediment to pursuing graduate work, where I would have the opportunity and freedom to pursue real research in the area of genetics. In time, the desire to get out of undergraduate work and into graduate school became an obsession, particularly since my "research" on my mutant held out was going well; I had prepared a short manuscript (with Rifenburgh as second author, at his request, Novitski and Rifenburgh 1937) and, generally speaking, was in a state of euphoria.

Actually during that first year Rifenburgh suggested that my projects involved questions that he was not well equipped to answer and that it might be a good idea to contact Calvin Bridges (Figure 2) at Caltech, an internationally known fly geneticist at who was undoubtedly the world's authority on the mutants of Drosophila melanogaster. This I did; he was extremely helpful and for a period of more than a year we exchanged letters at intervals of every few weeks, totaling about sixty pages. He, for instance, modified my designation for "held out" and made it "heldout."

At the beginning of my second year I applied for junior standing, much to the registrar's displeasure, and at the middle of my second year I filed for graduation status the following June. The registrar (not yet aware of my careful plans to knock over additional courses during the coming semester) was irate at my impudence.

Nevertheless, towards the end of my second year, despite strenuous efforts, I lacked about five credit hours necessary for graduation. The Dean of the School of Science (Howard Enders) offered me the needed hours on the grounds that I was doing extracurricular research on Drosophila and had published a small paper on my mutant "held out." This gift

He raises fruit flies by the millions

Research Laboratories

Dr. Calvin B. Bridges, famed California Institute of Technology biologist, studies fruit flies to determine how they—and hence humans—inherit their characteristics. "In their offspring, heredity and development of flies follow the same rules as in the children of the Smiths and the Browns," he says.

Keystone

Figure 2. My first letter to Calvin Bridges was sent in February of 1937, and his reply was dated 2 March 1937. Between that time and July 13 of 1938, a total of thirty six letters were exchanged, nineteen from me and seventeen from him to me, the latter being replies to my comments, inquiries, and especially detailed explanations of the specific steps involved in carrying out certain experiments. On 11 January 1938 I sent him a copy of a photograph of him which appeared in a college magazine, asking that he autograph it.

of credits came with one stipulation: that I not use any of the Drosophila work I did at Purdue as a basis for any thesis work for any advanced degree in the future. I discarded most of my Drosophila lines, sending a few (including Star-recessive, (Figures 3 and 4) later changed to Star-asteroid) to my high school buddy, Eddie Lewis, who had moved from a small college in Pennsylvania to the University of Minnesota. I urged him to look at the Minnesota graduation requirements to see if he too could find the loophole that would make it possible for him to shorten his undergraduate stay and then to join me at Caltech. This he did, and was able to cut a year off the normal four year undergraduate requirement. I also urged him to get in touch with Bridges, who might prove as helpful to him as he was to me.

At the time of graduation, it somehow became public knowledge that I had gotten my degree in two years and as it turned out, the university was giving an honorary degree to another graduate who had graduated in two years back in 1910. We got our picture taken together for the local newspaper. But I got into the newspaper in a different way too. Purdue, as an engineering school, had at the time no way of crediting those graduates who (presumably) had done a better than average job, with designations like "cum laude," etc. Instead students with high grades were graduated "With Distinction." When it came to time for me to cross the stage to pick up my diploma, the president (Edward Eliot, as I recall) was carried away by the spirit of the moment and announced that I was graduating "With *High* Distinction." After a sprinkling of applause (I had no relatives in Indiana at the time) the comment was forgotten. Or so I thought until the next day, when there appeared in the Lafayette newspaper a statement from Eliot to the effect that graduation honors at Purdue did not include the designation "With High Distinction," only "With Distinction." Thus was that early claim to fame abruptly torn

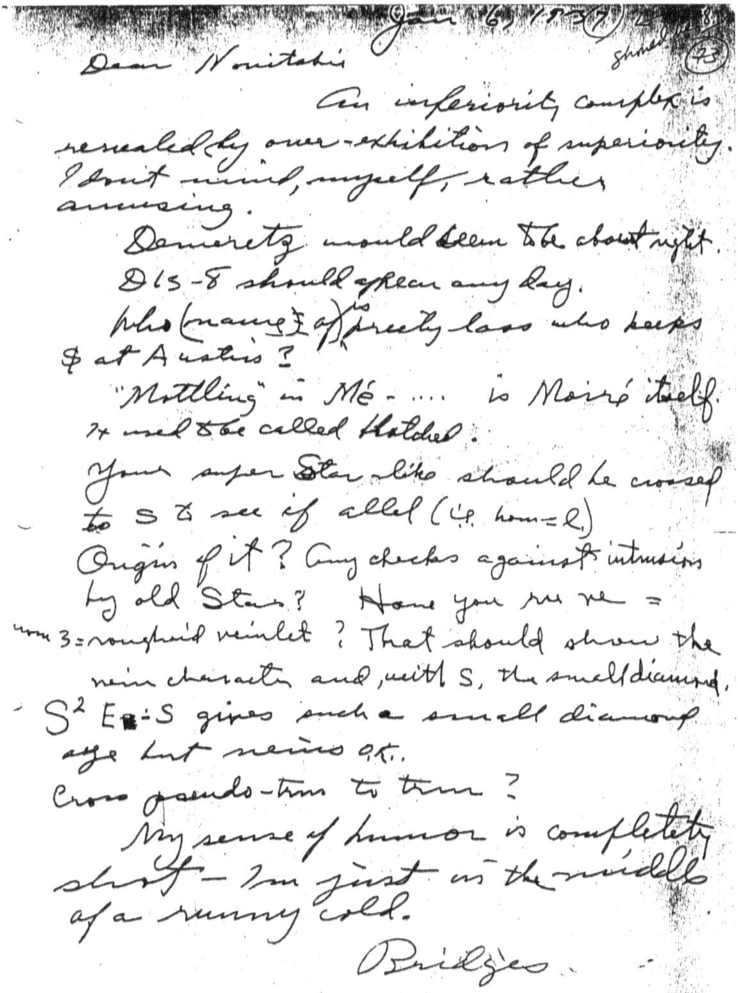

Figure 3. Facsimile of a page of a letter from Bridges to me dated 6 January 1938. In it he suggests the specific first step necessary in the analysis of the mutant I had found which greatly enhanced the roughening of the eye caused by the dominant Star.

character had never been called to my attention before because of my poor technique;
I do not examine F_1 offspring in great detail in most experiments and I have not had
much use for the inversion carrying Sb Me. Thought Moiré was a man's name.

Am having a protegé at Minnesota work on Star-like. Have explained to him
what should be done, and included an allel test with S, so you see I sometimes have
pretty good ideas too. Will know the results any day now.

Remember the vermilion allel I said I was going to discard. It was found
while deriving a b bw ♀. allel (v^{371}♀ ♂♂ x v ♀♀ = all v offspring. The vermilion
brown combination is white with a trace of pink, but the v^{371} b bw combination is
tannish or a lemon-color. Effect does not seem to be due to age differences. Does
b ever exert any effect on eye color? Do two allels ever produce different effects
in combination with another character even though they are exactly the same alone?
If your answer to both these questions is negative or indefinite, we have a pretty
little problem on our hands, one of the Ind Acad of Sci calibre.

I suppose you heard about Goldschmidt's "new" theory of chromosome structure. Who is going to have the privelage of squelching him? If you are going to do
the job, will you please send me a reprint? Willier of Chicago insists that he is
simply calling the whole situation by another name without making any great contribu
tion. Also, is it compatible with the fibro-crystal idea? Surely you are not going
to sit back and offer no opposition. I'll be terribly disappointed if you sit
back supinely and let somebody else do the job. It shouldn't be hard. Goldschmidt
messed up gametogenesis in the Trematodes and really didn't do such a brilliant
job in the interpretation of his results with Lymantria. The complete reasoning lea
ing to his conclusions is to be found in a book as yet unpublished. He says that
Muller came to the same conclusion but refused to back it because it was not in acco.
with present views regarding chrom structure. Smart man this Muller.

In a Cy/trm ♀ x trm ♂ cross, I am getting crossing-over. To the extent of
10/569 or 1.8%. Would it be best to keep trm as Cy, trm/Inv or 1 or Cy/trm?

Sincerely yours,

Edward Novitski

Figure. 4. A page of a letter from me to Bridges dated 11 January 1938, noting that I had sent the mutant to Eddie Lewis who had moved from Bucknell College to the University of Minnesota. I present the suggestion of a test for allelism given to Lewis as my own, although clearly it had originally come from Bridges. This test eventually gave unusual results and, after Lewis renamed the mutant "asteroid", it formed the basis for his Ph. D. thesis at Caltech and his subsequent concentration on the nature of activity of specific loci.

from my ego. Even now I have visions of multitudes of angry parents descending on the president's office demanding to know why their kids were not also graduating "With High Distinction." In any case, I can lay claim to the unique honor of being the only Purdue University student ever to have graduated with "High Distinction," on the assumption that such an honor, once awarded, would, like the Ph. D. degree, not be subject to revocation!

When the news of my "accomplishment" became generally known, so many students proceeded to file for credit by examination that the faculty quickly (and wisely) closed that unfortunate loophole. While the scheme that I had taken advantage of was effective in shortening the undergraduate period, it also had the effect of allowing a student to get his degree without any possibility that he or she might get a real college education. To this day, the one college course I now recall most frequently was one I was forced to take to fulfill a humanities requirement, a course on modern English and American poetry.

I met with Calvin Bridges very briefly one afternoon at Cold Spring Harbor during the summer of 1937. We discussed some genetics problems and in particular the matter of choice of graduate schools.

At Bridges suggestion, I applied during the spring of 1938 for financial assistance to several graduate schools. I applied to Caltech, Columbia, Berkeley, and the University of Texas. The first response was from Columbia, from the Zoology Department head there, one Franz Schrader (a very well-known cell microscopist). It was in the form of a telegram to Howard Enders, who was Biology Department head at Purdue as well as the Dean of the School of Science. The essence of this message was "Is Novitski Jewish?" (5-1). This was my first but certainly not the last brush with anti-Semitism in the academic world. However, this became a moot point because I shortly received a letter from T. H. Morgan offering me a teaching fellowship at Caltech amounting to $300 per year as well as room and board. I wrote to Columbia and Texas explaining that I had decided to go West.

At this point, Bridges suggested in a letter that I spend the summer at Cold Spring Harbor, where I would be appointed as a summer research assistant to work with him with some small stipend. As it turned out, he himself was not able to spend the summer there and he dropped by for a brief visit of a day or so at the end of summer—or so I was told. I never saw him that summer. I thought it most curious that after having spent so much time and effort in correspondence, he would not even bother to say hello.

Enfeebled by illness (a heart condition, I was told) Bridges returned to Pasadena. He showed up at the lab only very rarely. After I arrived in Pasadena I saw him on a few occasions as he waited for the freight elevator on the second floor of Kerckhoff where his lab was located and as I happened to be coming down the stairs from my lab on the third. When our eyes met, he would give a only a small smile and turn away, and in my introspective way I interpreted his unwillingness to be more friendly as a result of his disappointment at my forsaking work with Drosophila melanogaster for population work with other species. I did not know then what I do now, that his illness was terminal, and he knew it.

For this reason he may have been unwilling to pursue any kind of acquaintanceship for his own reasons related to his health. He died during the Christmas week of 1938 (5-2).

Summer of 1938 at Cold Spring Harbor

It is at this point, in the summer of 1938, that I entered the scene and became involved in advanced genetics research.

Although Calvin Bridges wrote suggesting that I spend the summer of 1938, after graduating from Purdue, working for him at Cold Spring Harbor where he would be spending that summer, he did not appear and instead I spent the summer there working as a student assistant with B. P. Kaufmann and M. Demerec analyzing the giant chromosomes of Drosophila melanogaster in larval offspring after the parent flies had been subjected to varying doses of X-rays. Dobzhansky was there also. When he and I learned of each other's presence and that we were both headed for Pasadena in the fall, we naturally fell together and, although we worked in different buildings, were practically inseparable outside of working hours. This duo was in fact a trio. Pio Koller, a former Dominican monk whose superiors had made the mistake of sending him off to England for a higher education, had just completed a fellowship at Caltech, where he had worked with Dobzhansky, and was now on his way back to Europe (6-1).

When it came time to leave to go west, Dobzhansky and I bought tickets on the same fast train, the El Capitan, which traveled between Chicago and Los Angeles. I spent much of my time reading his book, *Genetics and the Origin of Species*, which had just been published and which was receiving wide acclaim as a "masterpiece." I was delighted to find several trivial errors of which I thought he was unaware, first that the drawing of moving chromosomes on the cover had the chromosomes moving backwards, and second, that he had failed to include in his bibliography any citation to Darwin's classic, which, after all, had served as the basis for his title. These amusing oversights were corrected in the second edition (6-2).

During that summer and the subsequent train trip Dobzhansky proved to be an indefatigable conversationalist. He did not hesitate to discuss his colleagues at Caltech and, as a beginning student, I was flattered by what I considered to be his confidences. There were two who seemed to be treated gently in his comments, T. H. Morgan and Sturtevant.

Still, there were indications that all might not be going well in that group, although I realize in retrospect that I was not sufficiently sensitive to these nuances at the time. The first was his intonation when he referred to Morgan (who, as you will recall, instituted genetic work on the fruit fly back in 1910 and who was commonly accepted, in the US at least, as the Father of Genetics, second only to Gregor Mendel). Dobzhansky always referred to him as "the Great T. H. Morgan" with a slight but unmistakable emphasis on the word "Great" that made me feel that his use of the capital G even in the spoken word clearly carried with it a suggestion of disdain.

The second was a statement of his view of the future of genetics, the essence of which he repeated several times, which went something like this: "Sturt and Morgan have the opinion that the future of genetics rests in the application of the principles of physics and chemistry to genetics problems. I on the other hand believe that classical genetics, particularly the laboratory studies on the genetics of Drosophila melanogaster, is now dead, and the future will be spent most profitably on the study of natural populations."

In several respects this statement was quite to the point. The most basic phenomena of heredity, as revealed by the experiments with D. melanogaster, had already been examined in detail by the work of the preceding 28 years and it was becoming increasingly difficult to plan a program in this area that would inevitably lead to an acceptable Ph. D. thesis. Population work, on the other hand, was just beginning and the possibilities seemed infinite. Also, as I realized later, such purely observational work would superbly serve as a thesis problem because it could not possibly fail, unless, of course, the student was at fault.

The comment about the relation of physics and chemistry to genetics was anachronistic; Sturtevant's view was profoundly prophetic, but would have to wait for another twenty years to see its fruition.

One aspect of the relationship he did not mention was that Sturtevant had already expressed a lack of interest in work of this particular type of population study and had suggested that Dobzhansky not pursue it further (6-3).

When Dobzhansky suggested that I do my thesis work studying the distribution of lethal genes in a natural population, paralleling a work that he had just completed with his assistant (6-4), I readily agreed. It was not just a matter of being handed a thesis project on a silver platter, or even the prospect of spending weekends on field trips exploring the mountains of California. As one who had previously gone only slightly further west than the Wabash River in Indiana, these prospects certainly had their attractions.

But there was an additional problem to which his suggestion provided a solution.

In my effort to get into graduate work in a minimum of time, I had made a Faustian bargain with Howard Enders, Dean of the School of Science at Purdue. In return for five credit hours I desperately needed to complete my BA degree requirements in two years he made me promise I would not use my work with the Drosophila melanogaster (which was mostly, but not entirely, on the heldout mutant) and Dobzhansky's suggestion that I work on species other than melanogaster fit in nicely with that commitment.

The three day train trip afforded much time for conversation. One item still sticks in my mind. He recalled how one time when he went horseback riding he broke a leg when his horse, which was blind in one eye, trotted into a gatepost. Dobzhansky told, with a trace of emotion, how a student of his, Bob Helfer, put him into his car and, with great skill (avoiding sharp turns and sudden stops) drove him to the hospital. It was during this period of imposed physical inactivity that he collected together his lecture notes, and put them together in the book *Genetics and the Origin of Species.*

SECTION 3

HISTORICAL BACKGROUND

Sturtevant's Early Years

Sturtevant was born in Jacksonville, Illinois in 1891. He completed his undergraduate work at Columbia University, finishing the requirements for the bachelor's degree in just two years. During that time he published his first scientific paper, which was on the inheritance of coat color in horses. He received the degree of Doctor of Philosophy in 1914. During World War I he served as a private in the Medical Corps of the U.S. Army, stationed in the states. After that he returned to Columbia as research associate with the Carnegie Institution of Washington (7-1).

At Columbia, Sturtevant had become associated with Thomas Hunt Morgan, an embryologist of considerable reputation, whose interest at that time was diverted by the sudden appearance of white-eyed fruit flies (Drosophila melanogaster) in a culture of normal red-eyed flies that he was carrying along as a possible object of scientific research. The inheritance of this new eye color was worked out in detail, and its publication in the journal Science in 1910 marks the beginning of the science of genetics in the United States (Appendix A).

So, during those early years, research on the fruit fly proved to be a gold mine, and scarcely a year passed by without several significant discoveries on the fundamental nature of fruit fly inheritance. The relevance of this work to living organisms in general and to humans in particular was only slowly appreciated by most biologists (and not at all by nonbiologists) (7-2).

The students working at Columbia during those early years included several besides Sturtevant who would become legends in the field of genetics, notably H. J. Muller and C. B. Bridges, each of whom deserves serious attention in any historical account of the field.

In the late nineteen twenties, R. A. Millikan, the well-known physicist, had been empowered to transform the relatively unknown Throop Institute in Pasadena into a more advanced educational institution, the California Institute of Technology.

In 1928 Morgan was invited to form a biology department. He accepted and took the Drosophila group, notably Sturtevant and Bridges, with him from Columbia University. Sturtevant was given a faculty title and Bridges went as a Research Associate with the Carnegie Institution of Washington (7-3)

Dobzhansky's Early Years

Dobzhansky's background was quite different. He was born in the Ukraine in 1900 of parents who belonged to the Orthodox Church. His father's family was originally Roman Catholic, but converted as a matter of convenience when they migrated from Poland to the Ukraine. His father, as a teacher of mathematics, bequeathed to his son a Polish name but apparently none of his interest in or facility with mathematics. In his later years he remarked that he considered this to be a serious deficiency in his education. He attended the University of Kiev for four years starting in 1917 and in 1924 he moved to the University of Leningrad (7-4).

Dobzhansky's interests, training and subsequent orientation was that of a naturalist. In his memoirs he states that although he took university courses in physics and chemistry, his interests were such that he avoided as many of the classes as possible and concentrated on biology courses.

About his university training and lack of exposure to mathematics Dobzhansky says in his Oral History (p.33): "Biological or geological students had to have chemistry or physics alright but somewhat less stiff and unfortunately they did not have to have calculus. So, alas, I never had calculus or any higher mathematics which was and continues to be bad for me." As we will see later this can also be interpreted not as a deficiency in education so much as an exceptional development of the linguistic and naturalistic intelligences. We may wonder if sometimes one aspect of intelligence flowers at the expense of another.

In Russia, Dobzhansky had performed dissections to learn of the differences in internal anatomy of different strains of Drosophila, and had published work on the different color patterns exhibited by ladybird beetles. However, it was clear to him that to pursue more advanced work in the challenging field of genetics, he would profit from a close association with active workers in the field. So, in 1927, Dobzhansky came to the United States as a Rockefeller Fellow to study with the Morgan group at Columbia.

The Early Association of Dobzhansky with the Drosophila Group

Morgan had been given carte blanche by Millikan and the new Caltech administration to make several additional appointments in the new biology department, so he found it quite possible to add Dobzhansky, who was clearly, after all, a highly intelligent and promising prospect. The transition from the status of a visiting student to that of a permanent guest was not easy and Sturtevant was one of several who intervened on his behalf on several occasions to keep him from being returned to Russia. He was clearly Sturtevant's protégé.

Certainly the proposal of an appointment of someone without a Ph. D. degree (7-4) would have been subjected to the most intense scrutiny by the upper administration, but the advocacy by Morgan and Sturtevant, along with the necessity to provide evidence of permanent employment to satisfy immigration requirements, favored positive action. And there remained in the background the knowledge that Dobzhansky would almost certainly suffer a cruel fate if he were forced to return to Russia (7-5).

Dobzhansky's association with Bridges and Sturtevant proved exceptionally fruitful. After Muller's report in 1927 that X-rays broke chromosomes resulting in new combinations of chromosome segments, Dobzhansky published about a dozen papers centering on the genetic properties of such abnormal chromosomes. Because he was a novice in the art of Drosophila experimentation, which had already reached a high degree of complexity, it is certain that both Bridges and Sturtevant initially played a substantial role in the design and interpretation of those experiments, and Dobzhansky's acknowledgments in those papers indicate this to be the case. But most of Dobzhansky's scientific papers of that period dealt with two other aspects of Drosophila research, the detailed microscopic studies on chromosome structure and observations on the anatomy of different forms and genetic variations of Drosophila. In particular he became adept at the analysis of the giant chromosomes of the salivary glands of the larval stage (Appendix B).

In several respects Dobzhansky and Sturtevant complemented each other to their mutual benefit. First, Dobzhansky enjoyed looking through a high power microscope at the contents of cells; Sturtevant did not. A low power binocular for counting flies was his standard optical aid and he trusted the results of genetic experiments more than visual observation to reveal inner mechanisms (7-6).

In several of their joint publications, it was Sturtevant who presented the basic problem, as revealed by his genetic tests, followed by Dobzhansky's more extensive analysis of the visible form of the chromosomes involved.

In addition, Sturtevant's experimental procedure generally consisted of a rather small series of Drosophila matings, sufficient to make a point. Once established, that point could be expanded by Dobzhansky to utilize the full efforts of an assistant, as well as a good part of his own time. Thus, after Sturtevant, following the lead of a Russian worker, Chetverikov, who worked on a European species of Drosophila, satisfied himself that mutant genes abounded also in natural populations of Drosophila pseudoobscura and published a short paper to that effect (Sturtevant 1937), Dobzhansky published several longer papers in which the frequencies of lethal genes were determined for different localities in the mountains of Southern California (Dobzhansky and Queal 1938). When Sturtevant had the inspired thought that a certain class of chromosome changes found in natural populations must follow a specific set of serial steps in their origin, Dobzhansky provided the microscopic evidence that this was indeed the case, and followed it with long-term studies of those rearranged chromosomes from specimens captured from a large number of locations over a long period of time.

Another difference between the two lay in their laboratory approach to scientific problems and the publication of results. Sturtevant was phenomenon-oriented; he would try to recognize a problem and figure out how it might be approached, and would prefer to do the initial lab work single-handedly. If the results were positive or interesting, he might publish. Dobzhansky, on the other hand, was more data-oriented (7-7).

Sturtevant and Beadle

George Beadle was also part of the Drosophila group during the early 1930s. He got a Ph. D. from Cornell in corn genetics then changed his focus of interest somewhat. Although he collaborated with Sturtevant on several projects, two were outstanding.

The first of these was a phenomenal text on elementary genetics (Sturtevant and Beadle 1935). Unlike the extensive multi-volume texts common these days, this was short, to the point, and quite difficult to understand. It never did achieve great distribution. It had questions at the chapters' ends that even instructors could not answer (8-1). After the book's distribution, there went out an anguished cry for an answer book for instructors. This book exemplified the essence of the logical-mathematical intelligence, whereas at that time biologists as a group and biology teachers in particular were more likely to be represented by those in whom the naturalists intelligence was predominant. And it was no doubt here that Dobzhansky's great facility in the linguistic area and his weakness in the mathematical area must have become most evident to his colleagues, for it would inevitably become clear that this aspect of genetics was not his forte.

We have no direct evidence from the record of the time on this point, but we do have Dobzhansky's later description of those experimental procedures. Those described in the Sturtevant-Beadle text revealing the innermost workings of the meiotic process, are termed "tricky," a word whose implication of triviality is evident. A more sympathetic writer might have preferred words like "ingenious" or "brilliant." If it had not been obvious previously to the two that they had different mind-sets, inevitable discussions surrounding the contents and emphases in this book would certainly make it evident. Thus is not to suggest that Dobzhansky was incapable of mastering the advanced problems in that text, but simply that this kind of genetics did not represent his primary interest.

The second accomplishment of Sturtevant and Beadle centered on their work on inversions. If one were to ask a knowledgeable geneticist to name the two or three most important genetics papers of the decade of the thirties, the answer would almost certainly include the paper on crossing over in inversions in Drosophila melanogaster by Sturtevant and Beadle (1936). This paper presented the most detailed experimental approach to meiotic phenomena, but incidentally included the solution to one of the most puzzling problems in population genetics of that era: How were chromosomes with a section inverted from the normal order of genes able to survive in natural populations?

It should be noted that this formulation solved one of the outstanding problems of population genetics: why inversions which were subject to severe negative selection when in the heterozygous state were not eliminated from those populations in which they occurred.

The basic problem, shown in Figure 5, is that when a normal chromosome and one with an inversion try to pair, gene for gene, along their entire length, the chromosomes pair in inverse order, and when exchange occurs between two of the four chromatids during germ cell formation (meiosis), two inviable types are produced, one with two centromere regions and the other with none, and both of these have an irregular (and usually inviable) assortment of genetic material. If they should be delivered to the egg, they would cause the death of any developing eggs carrying them. This loss (of one normal and one inverted chromosome) would not be equally felt by both types; the one with the lower frequency would be most adversely affected and this would lead to its quick and almost certain elimination from the population (Appendix C).

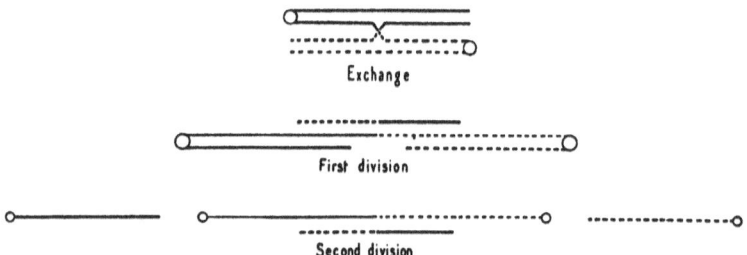

Figure 5. Results from crossing over in heterozygous inversions. Schematic figure from Sturtevant and Beadle (1936) with their explanation: Single exchange within a heterozygous inversion. The upper figure represents the two X's of a female in which one chromosome is practically wholly inverted. At the first meiotic division there results a chromatid tie; this leads to an orientation of the second division such that the two terminal nuclei receive only non-crossover chromatids; one of these is the egg nucleus. The result is the total loss of all single crossover chromatids to the polar body nuclei.

The first problem that those workers faced was to determine conclusively that the necessary initial event, the exchange of segments within the relatively inverted regions, actually did occur. This they did by constructing a special chromosome (technically a tandem metacentric), which was, for all practical purposes, an inversion heterozygote but one that produced viable products (Figure 6). Not only were predictable rings produced by crossing over within the inverted sequences, but they were recovered with a higher frequency (8-2) than expected!

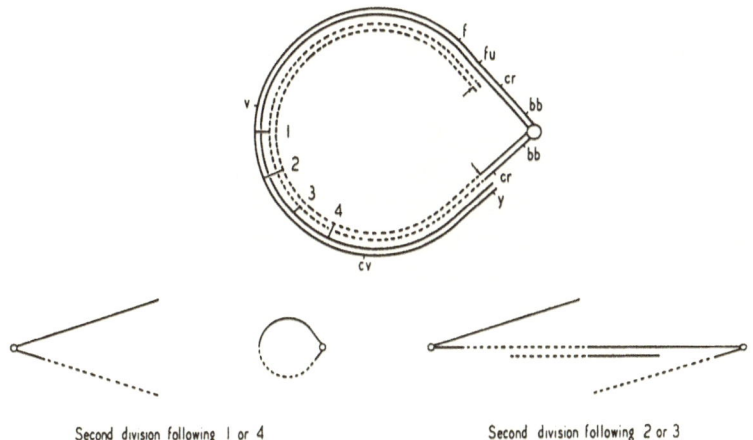

Second division following 1 or 4   Second division following 2 or 3

Figure 6. Results from crossing over in heterozygous inversions where the two chromosomes are attached together to form a single unit, technically known as the tandem metacentric chromosome. Chromatids resulting from single exchange are shown at the lower level. The ring chromosome, second from the left, is the viable product after crossing over within the equivalent of a heterozygous inversion complex. It is the excessive production of such rings that confounded Sturtevant and Beadle (1936) as well as Sidorov, Socolov and Trofimov (1935). These were later shown to be recovered in excess because of a non-random orientation of the centromere regions at the second meiotic division and was called "nonrandom disjunction" (Novitski 1951). It provided the main justification for proposing the phrase "meiotic drive" as a catchall for those cases where gene frequencies can be altered because of some departure from normal segregation rules at or about the time of meiosis, independent to some degree from the ordinary concepts of Darwinian selection (Sandler and Novitski 1957).

The second problem centered on the question of why the abnormal chromosomes produced after crossing over in simple inversion heterozygotes did not kill the eggs that they got into. After many tests, they concluded that this was the result of the preferential inclusion of the inviable products in the polar bodies, where their loss would be unnoticeable. This conclusion had been suggested also by work of B. McClintock on corn, and was not verified cytologically for animals until many years later.

Thus these workers solved one of the most perplexing problems of population genetics of that era. However, if you look at the "historical" accounts of the development of the study of population genetics covering that period, you would be hard pressed to find any mention of the significance of this work. (I have not found any in accounts written by historians of science.) Perhaps the main reason is that the experiments are described in technical terminology that is comprehensible only to those trained in that area. Another is the characteristic of that group of historians, all of whom possess an extraordinary facility with the English language, to accept more readily those aspects of biological endeavor that they can easily understand, and apparently, the more elegant the language of presentation, the more acceptable their view of the scientific concepts. Further, when faced with an incomprehensible scientific construction, or with a paper incompatible with their view of what constitutes meaningful science, they are likely, consciously or not, to take refuge in a rationalization that the work is therefore unimportant.

This minimization can be seen, for instance, in the use of the word "tricky," which suggests insignificance, a word that was used by Dobzhansky and by historians of science following him. That this word would be used to describe such simple genetic constructions as attached X's, triploids, etc., all of which played important roles in developing our understanding of basic biological mechanisms, would confound most geneticists.

What genetically trained worker would not marvel at the way that genetic experiments with the attached X chromosome immediately proved that crossing over occurred in the four strand stage of meiosis, and showed that the centromere region, that part of the chromosome involved in chromosome movement at the time of cell division, was located at the end of the chromosome that carried the gene for carnation eye color, and not the other end, which carried the gene for yellow body color? And all this without any visual evidence in Drosophila that this was so!

A comparison of the work on inversions, both investigated by Sturtevant, one on D. melanogaster with Beadle and the other on D. pseudoobscura with Dobzhansky, is instructive because it reveals the fundamental differences of the workers in their approaches to scientific problems, differences that may have played a major part in the dissolution of their scientific partnership as well as the departure of Dobzhansky from Caltech.

The inversion work with pseudoobscura indicating phylogenies was instigated by Sturtevant's realization that overlapping inversions provided an unambiguous picture of the order in which they originated (Sturtevant and Dobzhansky 1936) and made it possible for Dobzhansky (Dobzhansky and Sturtevant 1936) to construct a phylogenetic tree (8-3) using the giant polytene chromosomes. The result was an impressive chart showing the path that the various chromosome alterations must have followed in their history. But these results pertain to the third chromosome of D. pseudoobscura only and, although the original idea has an aspect of universality, the final elaboration does not.

On the other hand, the work with melanogaster starts with questions that demand answers. One of these is: why are these chromosome alterations (inversions) that predictably should stand at a considerable disadvantage and so be eliminated from populations, sometimes found in abundance? Sturtevant and

Beadle present the answer, after going through a series of logical steps to validate their conclusions. As their discussion makes clear, their results have applicability not only to Drosophila melanogaster and not only to insects but to animals and plants in general. This represents a fundamental difference between work on D. melanogaster and D. pseudoobscura sometimes not recognized by those who regard the two species as two equal competitors. The work with melanogaster generally had a universal aspect, the conclusions usually applying to diploid plants and animals in general, while that with pseudoobscura was more inclined to be species specific.

Development of a Rift

The first indication that perhaps not all was going well between Sturtevant and Dobzhansky came after Sturtevant returned from a sabbatical year in England in 1933. It appears that the warm and friendly association that had characterized their relationship previously had turned into a colder and more distant one. Dobzhansky's interpretation was: "After having spent a year in England he came back a very different person. He just was not the same man. However bitter that situation was and is to me, perhaps it did a vastly greater harm to him. Since then, and that is consequently more than a quarter of a century, he did very little in the way of research and he's published very little research." The logic of this statement seems a bit strained, but it does reveal Dobzhansky's preoccupation with publication. This does, after all, cover the period when Sturtevant showed that wild populations of D. pseudoobscura carried lethal genes, and that the different inversions in that species could be arranged in a phylogenetic sequence. Both of these formed the bases for much of Dobzhansky's later work. This was also the period that included the classic work of Sturtevant and Beadle on inversions, one that had profound implications in population genetics. In addition, Sturtevant completed a monumental study on nonrandom chromosome segregation, a study whose implications to population genetics are only now beginning to be recognized.

Sturtevant, not inclined to discuss personal matters, never revealed the reason for his change of attitude and various guesses have been made. One is that Sturtevant's apparent Anglophilia, developed during his sabbatical stay in England, alienated Dobzhansky (although why this should have such an effect on Sturtevant's attitude towards Dobzhansky rather than the reverse is not at all clear). Another is that Dobzhansky was proselytizing incoming graduate students in Sturtevant's absence, but an examination of records of the few graduate students during those early days does not suggest any disproportionality, and no proselytization (at least none before I arrived at Cold Spring Harbor). Still another suggestion is that Dobzhansky's work with D. pseudoobscura antagonized Sturtevant, but the emphasis on pseudoobscura came later and it was shared jointly by the two. Other circumstances may have played a role. Possibly Sturtevant became aware of Dobzhansky's religiosity, developing or dormant, something which Sturtevant would find intolerable, quite unacceptable, in a first-rate scientist. He may have somehow sensed Dobzhansky's increasing alienation from the kind of genetics research demanding a logical-mathematical bent, practiced by Sturtevant and most others in the field, and Dobzhansky's preference for investigations with a more naturalistic flavor.

However, none of these, it seems to me, would evoke the strong negativity that Sturtevant seemed to be expressing at that time. I myself have a guess (and it is only a guess, but I think that it is more likely than others that have been proposed). Unfortunately, it is not based on any scientific consideration, but on a more personal one.

Dobzhansky was something of a raconteur. He enjoyed the attention of his audience when telling stories and his audience enjoyed listening. Unfortunately, these stories sometimes included tales that reflected adversely on other people. For instance, in his Oral History Memoir on p. 636 he writes about a worker at Oak Ridge National Laboratory: "Curiously enough, not only [here I have deleted the name] who, since I believe this interview will not be read for some time, I may say, in my opinion, is a rather stupid man, but even many other people who are not stupid, simply fail to see the need of investigating the manifestation of the mutations in living populations."

Besides Sturtevant, Bridges was the one other person in the lab to whom Dobzhansky owed a debt of gratitude for his constant and patient tutoring in the art of Drosophila experimentation. Yet Dobzhansky did not hesitate to discuss, with relative strangers and in a joking way, details of Bridges' romantic affairs. I heard these first from Dobzhansky in 1938 in the mess hall at the Cold Spring Harbor labs when the stories came out in his characteristic high pitched voice, meant for all to hear (to the great amusement of the audience). I, as a beginning graduate student, laughed with all the others and it was only much later that I wondered if this sort of storytelling was discreet. Dobzhansky states in his oral history that these affairs were common knowledge. Perhaps, but it would appear that he himself might have been responsible to some extent for their proliferation.

It is my guess that the cause for the initial estrangement of the two was not related directly to Sturtevant's stay in England. On the way to his sabbatical in England, Sturtevant went to an International Congress of Genetics in Ithaca, N. Y. There he and Dobzhansky presented a joint paper (Dobzhansky and Sturtevant 1932). At such meetings it is customary for small groups to aggregate during the evening hours to engage in discussions, partly on the more interesting papers of the day, but also to gossip, much as all small groups of humans, both male and female, do. Here Dobzhansky might very well have made comments about his colleagues, possibly Morgan but particularly Bridges, similar to the story I heard him tell about Bridges at Cold Spring Harbor. Inevitably Sturtevant would have become aware of these, either at that meeting or later in England, and undoubtedly would have resented the disrespect being shown to a good friend and colleague he had worked with for so many years.

Another indication that all might not be completely harmonious between the two occurred somewhat later. For several years they occupied the same laboratory space, but one night Sturtevant returned and removed all his belongings and equipment to another. Dobzhansky attributed this move to his having made some uncomplimentary remarks about Sturtevant's mentor T. H. Morgan, to which Sturtevant apparently took offense.

Dobzhansky says in his Oral History Memoir (p. 273): "And so, one day we were sitting as always in our laboratory room with Sturtevant, and I made some critical remark about Morgan. I do not remember exactly what I said, but I am absolutely certain that it was a remark which was much less critical and certainly nowhere near as disparaging as hundreds or thousands of remarks which he used to make almost every day. And here came a reaction which startled me to an extent which probably left me with my mouth open for some time, namely, Sturtevant raised his voice and said: 'I ask you never again to make in my presence any disparaging remarks about Morgan.' I was left completely speechless. And that was the end of our friendship. Now, I say, only some sort of psychoanalyst can get at the bottom of that affair. Was it possibly that Morgan's act of generosity [in sharing the money from the Nobel Prize with Sturtevant and Bridges' children] made him ashamed of himself? Was it possibly the fact that I had been for years his confidante, having heard his innumerable disparaging remarks about Morgan? Maybe, under the circumstances, a person becomes unacceptable . . ." Once again it seems a bit difficult to follow Dobzhansky's line of reasoning. But there may be another interpretation, which is that Sturtevant was not reacting only to a specific comment made about Morgan, but more generally to the tendency of Dobzhansky to engage, on occasion, in personal remarks of a negative nature. And this would hit home when the object of the remark was a member of the original Drosophila group, to which Sturtevant would feel strong personal loyalty after so many years of association, but one to which Dobzhansky had less reason for such consideration.

The collaboration between the two came to a standstill sometime in 1936-1937. True, Sturtevant's demonstration of the precision of the sequence of succession of chromosome breaks in inversion formation was published jointly with Dobzhansky, but if the account of that "collaboration," told to me by Jack Schultz, is correct, it came about in an unusual way.

One day Schultz was visiting Sturtevant in his office and the latter, with paper and pencil, was explaining to Schultz why overlapping inversions provided a unique opportunity to describe their origin in a phylogenetic tree. At one point, Dobzhansky joined the two and listened, apparently without too much interest. But the next day, once again as Sturtevant and Schultz were talking, Dobzhansky appeared, and, interrupting the conversation, said: "Sturt, I have a suggestion. Why don't we publish on this idea of yours and I will do the chromosome details." This they did. It was eventually published as Sturtevant and Dobzhansky (1936). Some time thereafter, Dobzhansky published a more detailed chromosome tree, and it was published as Dobzhansky and Sturtevant (1938). Sturtevant's contribution to this second paper was negligible and his name was on it primarily as an acknowledgment to his contribution of the original idea. To refer to these papers as the work of "Dobzhansky and Sturtevant," however, (as some historians have done) is to reveal a personal preference on the relative values of the contribution of the initial concept vs. its detailed demonstration.

The final break between the two may have arisen in a somewhat unusual way; perhaps it was contributed to by a couple of Roman numerals that appeared in Dobzhansky's papers as they finally appeared in 1938.

We have previously discussed the different points of view of the two with regard to the kind of research they preferred. We are hypothesizing that these differences were determined by their mind-sets, that their orientations were determined largely by their intelligences, and that Sturtevant had a logical-mathematical bent whereas Dobzhansky was primarily naturalistic, with a strong linguistic component.

Sturtevant, after reading of the Russian work that demonstrated the existence of mutant genes in wild populations of European Drosophila, performed the same tests on Drosophila pseudoobscura and satisfied himself that such genes were found in American species as well.

On p. 415 of his Oral History Dobzhansky states, when discussing his projects to make collections of D. pseudoobscura in a number of different locations in the Death Valley region for comparisons of their genetic constitution: "I regret to say that . . . . unpleasant but it is a fact, I am supposed to be truthful in this interview . . . Sturtevant did everything in his power to discourage this work. He argued that the technique of detection of this concealed variability was not valid, and that the thing has no particular interest anyway."

In the spring of 1938 Dobzhansky published, with his assistant Marion Queal as co-author, a paper whose title began "Genetics of Natural Populations I." And in the autumn of that year came "Genetics of Natural Populations II." Thus, he was entering with full force into that area which, according to Dobzhansky, was "of no particular interest" to Sturtevant and further, by designating each title with a Roman numeral, signaled his intention to continue, presumably indefinitely.

Again, on p. 404: "I started to do it with geographic strains, in a way in which Sturtevant was not interested, in fact did not approve." This is the point alluded to above—that Sturtevant had essentially disapproved of the general trend of the thesis problem Dobzhansky suggested that I work on, long before that curious meeting in the basement of Kerckhoff. Only decades later when I read the Oral History Memoir did I understand why Sturtevant was so noncommittal during our initial meeting, and was hurling chunks of ice against the wall with unusual vigor.

From Sturtevant's point of view I was an incoming graduate student, presumably of some academic merit, who had been prematurely proselytized by Dobzhansky to work on his newfound interest, one of which Sturtevant disapproved.

From Dobzhansky's perspective, I was to be the first of a series of graduate students who would enthusiastically collaborate in the analysis of wild populations. Up to this point, the only worker besides his assistant Marian Queal to be involved in this work was Pio Koller, who had been in this country temporarily on a fellowship, but who had gone back to Europe.

My attitude at the time was one of excitement that I had finally achieved the status of a Ph. D. candidate who now could devote his time fully to the intellectual pursuit of fundamental work in genetics. What I did not realize was that the partition of my intelligences had left me inadequate to meet the challenges I would be facing.

SECTION 4

CAREER AT CALTECH

Graduate Work

My Initial Research Project

During the first year and a half we made many (perhaps a dozen or so) field trips into the mountains of California collecting D. pseudoobscura. On one occasion the two of us were augmented by a small group; the obligatory photograph at the top of Telescope Peak appears elsewhere in this book.

Dobzhansky was an avid naturalist and would point out to me many of the wonders of the California flora, all of which I listened to politely for I had no interest whatsoever in that aspect of biology that consists of naming things and memorizing those names, the consequence no doubt of a deficient naturalist's intelligence. He would try to impress me with the beauty of the color of the blossoming flowers, something I could not only not appreciate but could not even see because of my red green color deficiency (10-1). I was also a bad companion from another point of view. He would always drive and, overpowered by the heat of the day and the monotony of the trip, I would regularly fall fast asleep, which did not enhance my desirability as a traveling companion.

The research project Dobzhansky had outlined for me was, at the beginning, a challenge. I had to develop several new lines of Drosophila pseudoobscura with X-ray induced changes in both chromosomes numbered two and four, twice as many as had been done by Dobzhansky and Queal. Then it was a question of testing each new lethal gene I found against all others I had previously found, and as the number of flies Dobzhansky and I collected in the San Jacinto mountains accumulated the amount of routine work became unsustainable.

My enthusiasm for this work began to steadily diminish, not only because of the increasing workload, but also because I realized that after this routine work, which would last several years, I would be in the possession of nothing more than a couple of numbers around which I would be expected to fabricate a thesis of some length, coming to some conclusions of lasting significance in the field of evolutionary biology. And it would make not the slightest bit of difference what those numbers were, because I would be expected to then find the right words to justify the numbers. This was clearly beyond my literary capacity.

But most disheartening was the realization that the work I was doing, although arduous, was not intellectually challenging. In fact, I felt that my "research work" could just as easily be carried out by a well-trained chimpanzee. I had entered graduate work with lofty thoughts about creative research and applying my intelligence (to whatever extent I could) in tackling genetic problems. This project was a severe disappointment.

Tea Time

Every afternoon at about three thirty the staff and graduate students gathered together for tea in an otherwise unused lab at the end of the hall on the third floor. The staff included as regulars Dobzhansky, Sturtevant and Emerson. Occasionally Jack Schultz would emerge from the second floor to join the group. Very rarely, Albert Tyler, an embryologist from down the hall, would also join in. During that first year the graduate students who attended were Dwight Miller, Bob MacKnight, Bob Helfer, and me.

These meetings were purely social in nature. Topics generally included the weather, politics, local events, trips, national news and occasionally one of the staff would bring up a matter of scientific interest gleaned from a recent publication. One subject that never seemed to come up was religion. Dobzhansky's religious fervor (described later by others) was never in evidence, neither in laboratory discussions nor during field trips, which always occurred over weekends when one might expect such an attitude to be most evident.

Research problems were never discussed. There was a large portable blackboard at one side of the room but this was almost never used. During my four year stay as a graduate student, not once did I hear, at tea-time, a serious thought that I might have called a bright idea, a flash of inspiration, or an animated discussion of any person's research data (10-2).

I know that at times such good ideas would occur but exposition of these was held in small one-on-one discussions in the laboratory. It could be that the lack of communication about scientific matters may have stemmed in part from a fundamental disagreement between Sturtevant and Dobzhansky about the importance of some of its most basic aspects.

An exception to this routine came whenever a visitor dropped in, which happened every few months. This was before the days of easy accessibility by air, or ready funds to pay for guest speakers (a situation that changed drastically after the war when federal funds became available for grant-supported research). These visitors (and I remember Ernst Mayr and C. H. Waddington particularly) did not give formal seminars as became the custom later on, but would hold forth during tea-time. The discussion would involve the staff and the visitors, with the students almost always respectfully silent.

These tea-time discussions would develop into debates, in which the staff would question the visitor rather closely, and I was struck by the fact that the visitors always seemed to have the ability not only to defend their points of view but on occasion to outdo the staff members. But then the visitors were always on their own research territory and could argue from the vantage point of their own data and presumably they had had many previous discussions and hours of thought on their problems.

My Future Wife

During that first year, Dobzhansky called me to his office to meet a college student, Esther Ellen Rudkin, whose family he had known for years. Her brother George had been a student at Caltech and, as an undergraduate, had worked with Dobzhansky on several projects that were later published as *The Genetics of Natural Populations I and II.* The father Charles was an avid butterfly collector and had gone on a number of collecting trips with Dobzhansky, he to collect butterflies while Dobzhansky collected Drosophila. Dobzhansky and his wife Natasha were frequent dinner guests at the Rudkin home; some of these I also attended.

Esther Ellen had been ill and had missed the lectures on the principles of genetics in her college biology course. She had come to Dobzhansky for help and he suggested to her that it might be better if I did the tutoring. What he had in mind of course was something entirely different, and in fact it turned out as I believe he had planned. We were married four years later when I was in the army during WWII.

Graduate Programs of Others

In the meantime I learned about the other students' research problems.

Bob MacKnight, Sturtevant's student, was working on the hybrids between Drosophila pseudoobscura and miranda in an effort to discover why these two closely related species seemed to differ so radically in their chromosome makeup. (Dobzhansky had reported that one of them, miranda, lacked a chromosome present in the other.) Bob Helfer, Dobzhansky's student, was investigating the sensitivity of D. pseudoobscura chromosomes when exposed to X-rays. Dwight Miller, another of Sturtevant's students, was investigating the chromosomes of Drosophila algonquin, an eastern relative of the western Drosophila azteca, for which Dobzhansky and Socolov (1936) had reported an unusual chromosome complement. Klaus Mampell, a transfer student from UCLA, brought with him a most intriguing phenomenon involving the nonchromosomal

inheritance of a factor in pseudoobscura that elevated mutation rates considerably. William Hovanitz came to the
department with a preconceived and active program centered
on field work with the distribution of color variants in the cabbage butterfly, Colias. And I, as Dobzhansky's latest acquisition,
was to work on lethal genes in pseudoobscura. I was the first
(and only) graduate student to become part of the group
(which then consisted only of Dobzhansky and Marian Queal,
his assistant) concerned with the detailed analysis of D.
pseudoobscura populations.

As it turned out, MacKnight concluded that Dobzhansky,
in reporting that a chromosome was missing in Drosophila
miranda, had overlooked a more complicated and interesting
phenomenon in which a chromosome had become attached
to a sex chromosome, the Y-chromosome, and in that protected
state had started to degenerate. Furthermore, he challenged
Dobzhansky's observation that there were as many as six different translocations, interchanges between unrelated
chromosomes, that had occurred since the two species had
evolved from a common ancestor. In his several years of thesis
work he had seen no evidence of even a single case (MacKnight
1939). He was openly critical of Dobzhansky's work. I happened
to be present by chance during one altercation in which he
rather violently (and in my opinion somewhat unjustly) argued
with Dobzhansky, accusing him of incompetent work and challenging him to produce the material that led him to his
conclusions. Dobzhansky of course could not, and MacKnight
knew this: most of Dobzhansky's microscope preparations were
"temporary" ones, which were easier to make and were re-
puted to show more detail than more permanent preparations.
Unfortunately these temporary preparations always disinte-
grated in a relatively short period of time.

Dobzhansky himself admits in a rather indirect way in his
Oral History Memoir that he was wrong. At that point where he
briefly mentions this work, the word "translocation", which is
the main focus of the papers referred to above, is not men-

tioned at all. Instead he refers to these observations as "upsets." In his Oral History Memoir, p. 363 he says "We had eventually tried to make an estimate of how many breaks, inversions, and other upsets you have to assume to derive chromosomes of one species from the other." The word "translocation" has been replaced by "upset," a word which has no meaning in the language of the ordinary geneticist.

Dwight Miller somewhat later investigated the chromosomes of D. algonquin and indicated that this species, closely related to azteca, did have the chromosome arm reported by Dobzhansky and Socolov as "missing" in azteca (Miller 1939). This made it much more likely that their conclusion about translocations also was faulty.

I Get the Sack as a Teaching Assistant

As the second year of graduate work proceeded, I became more and more disenchanted with the thesis project that Dobzhansky had set me out on, and I desperately sought a way out. Under ordinary circumstances I might have discussed this with Sturtevant, but he was away at Harvard. I knew that I could not discuss abandoning the problem with Dobzhansky because he would most certainly apply extreme pressure for me to continue. And I could hardly tell him what I now thought of the problem, because this would be a direct personal insult to him. In desperation, I simply took the cultures I had been working on and put them in the hall where they would be picked up and washed by our food maker.

I was so tied up in my own miserable situation that I did not realize that doing this was also a direct affront to him, even though I knew that he, like most scientists, had a tender ego (11-1).

Without any problem to work on, or anyone from whom I might get advice, I decided that I would have to find a suitable research problem on my own. After a little thought I decided that I would try to find new cases where closely related species produced offspring, because an analysis of any peculiarities appearing in the offspring might give some insight into the mechanisms by which their common ancestor diverged into distinctly different forms. After many trials, I hit upon one interesting case, involving Drosophila repleta and its close relative Drosophila neorepleta.

This work was interrupted by a traumatic event. The annual meeting of the biology staff to consider appointments of incoming graduate students for teaching assistantships decided to make one additional one available by removing mine. My first thought was that this move was promoted by Dobzhansky, who had good reason to doubt my qualifications to carry through a research project required for an advanced degree.

According to one staff member, James Bonner (in plant physiology), Dobzhansky was my advocate, and had in fact argued in vain for my retention. This convinced me at the time that if anyone other than myself were to be considered responsible for this drastic blow to my ego, it might have been Sturtevant. Perhaps he had communicated his unhappiness with my performance on his final exam during my first year to other staff members, who, in his absence, carried out what they might have considered to be his wishes.

Another version of the events surrounding my nonreappointment suggests the opposite of the account given above. Ed Lewis, who was then a first-year graduate student, now recalls that Dobzhansky told him prior to the staff meeting that I would lose my teaching assistantship, which suggests that he was not my advocate as Bonner had described him to me and that Dobzhansky was actually the instigator of this move. If this is so, and I have no reason to doubt it now, many years later, then Sturtevant's action in appointing me his research assistant for the following year, effectively nullified Dobzhansky's wish to terminate my graduate career. It implicitly asserted Sturtevant's concern, and ultimate responsibility, for the welfare of the graduate students in the department, and must have put

Dobzhansky in an awkward position, which may have added to the pressures that caused him to leave for Columbia University. However at the time I was unaware of these possible complications (11-2).

Under the circumstances I might have quit and looked around for another school to continue my graduate work. I thought, for instance, that I might be accepted by the University of Texas where the department head, J. T. Patterson, had earlier offered me financial help and had sent me a rather long, very kind, and understanding letter after I wrote to tell him that I had decided to go to Caltech. There were however several circumstances which forced me to stay if at all possible, besides the natural reluctance to admit defeat in a situation where I felt I had been not completely to blame. One was that I was that I had been working under C. A. G. Wiersma, an animal physiologist, on a small research problem, centered on the nerves going to the heart of the crayfish and had hit upon a rather neat experiment which demonstrated conclusively that a chemical substance was produced at the site of the accelerator nerve. Wiersma guessed that it was acetylcholine (subsequent work by others proved him right). This work was subsequently published in the Journal of Experimental Biology (Wiersma and Novitski 1942). Curiously the best scientific work that I did as a Ph. D. candidate specializing in genetics was in the field of invertebrate physiology.

The most important reason for staying was that by this time I had become completely enamored of Esther Ellen Rudkin (to whom Dobzhansky had introduced me), and I was not willing to abandon my romantic interests to other competitors. Her parents were not at all pleased with the prospect of her interrupting her college career for the purpose of following me to Texas, or wherever I might go, which at that stage she would have been willing to do.

For several weeks matters for me were in turmoil. James Bonner discussed with me my improvised research problem on hybridization in the Drosophila repleta group and was not impressed. I would never have dreamed that the relationship of Bonner and me would be completely reversed some years later (11-4). Now I point out with some satisfaction that these hybridization observations provided the basis for a paper published by Sturtevant in 1946, in which he credits me for the initial observations. When Sturtevant broke the impasse by writing to me from Harvard (where he was visiting temporarily) that he would take me on as his research assistant the following year, the crisis was resolved.

My Life as a Research Assistant

Sturtevant suggested that as a research project I work on the giant salivary chromosomes (see Appendix B) of a species called Drosophila athabasca. This species was so closely related to Drosophila azteca (which Dobzhansky and Socolov claimed was missing a chromosome arm when compared with other closely related species) that they would mate and produce offspring in crosses made in the laboratory, although those offspring were physically abnormal with both sexes being completely sterile. At the time I did not question or even wonder about the propriety of working on a form so closely related to one that Dobzhansky worked on, with results about which some serious questions had been raised.

Drosophila athabasca proved to be, in my hands, a rather intractable object for chromosome studies. Others in the lab, particularly Dwight Miller, who had remarkable success with Drosophila algonquin, a more distantly related species, also had difficulty making reasonable microscope preparations of athabasca, so it was not entirely a matter of personal incompetence. However, in a relatively short period of time I was able to establish that the chromosome arm Dobzhansky and Socolov (1936) reported as "missing" in azteca (12-1) was present in athabasca (Novitski 1946).

The next step was to examine the chromosomes of the hybrids of the two species, azteca and athabasca. Here the corresponding chromosomes of the two species could be seen lying side by side, proving unambiguously the existence of the Dobzhansky's "missing" chromosome arm (see Appendix B). When I showed the microscope preparation to Dobzhansky, he seemed to me to be quite unhappy. What I did not know, and he did not tell me, is that he must have already seen this configuration and with H. Bauer had already published a brief abstract to this effect (Bauer and Dobzhansky 1936). I became aware of this many years later. Why they did not follow up this observation with a regular scientific paper of some length, the usual custom for most workers, and obligatory for Dobzhansky, is a matter for conjecture.

Duties as Research Assistant

In evaluating Sturtevant's reactions to Dobzhansky's observations suggesting the occurrence of translocations, an additional aspect should be mentioned. The position of "research assistant" to Sturtevant that I assumed in the fall of 1940 was a sinecure. Except for the routine care of his fairly extensive assortment of stocks of miscellaneous species collected throughout the United States, a core which took no more than a few hours a week, he did not need or even want any assistance. The only time I actually became involved in an experiment happened when he was in the middle of an experiment and had to leave town. He asked me to make some counts of flies each day of his absence, which lasted about a week. Eventually he published a short paper and attached my name as a co-author (Sturtevant and Novitski 1941a). I read the paper with some interest because up to that point I had not the slightest idea what the experiment was all about.

Unlike Dobzhansky, who, with his assistant, ordered quantities of culture medium in half pint bottles which, after use, were deposited in the hall daily to be picked up by the bottle washer, Sturtevant would ordinarily use a few small clean vials, perhaps a dozen or so, and into these he would put a small wedge of solid culture medium. At one point I noticed that he was carefully washing each individual vial at his laboratory sink and, feeling that this chore should be more appropriately be done by the hired help, namely me, I volunteered to wash them for him. He brusquely pointed out that he was quite capable of handling this chore himself, thank you.

Thus it was that the "assistant" did little assisting and was able to spend most of his time on his thesis problem. But the disproportionality of free to working time was painfully evident, even to Sturtevant. So after I spent a few weeks in this job, he did give me a chore to occupy more of my time.

One afternoon Sturtevant came into my lab (a small room adjacent to his equally small one) with a journal in his hand and made the following comment: "Now that Muller has finally published, I think that we can go ahead." This was a clear admission that, whatever the topic was, Muller had some sort of claim to priority. Sturtevant left for a moment and returned with a stack about six inches high of lined notebooks.

It turned out that these contained notes collected over many years, excerpts of publications throughout the world written by many dozens of different workers, describing the mutants that had appeared in many different species of Drosophila (12-2).

This was a project identical in aim to the translocation work of Dobzhansky, that is, it was concerned with the general integrity of the chromosome structure throughout the many species of Drosophila. Surely Dobzhansky must have been aware that at the same time that he was reporting the interesting (but incorrect) observations on the prevalence of gross chromosomal changes, translocations, distinguishing closely related species, Sturtevant was, and had been for years, collecting genetic information on this point that suggested the contrary.

Comparing mutant genes from one species to another is pretty crude and fraught with hazards, but it was all that could be done at that time when trying to find similarities between species so far apart that they do not hybridize. It is to be expected that future systematic studies made of similarities based on DNA hybridization techniques will lead to greater insights into the nature of species divergences over time.

So I was given the job of taking the information that Sturtevant had collected, along with any additional information I might get from further library research. I did not get much that Sturtevant had not already found, although I do recall my delight at finding a pertinent paper, written entirely in Japanese (which of course I do not know) that included the notation "w,1," a symbolism that meant that the mutant for white eyes (w) in that species was found on the sex chromosome (1), all that I needed to know to include it in our comparative studies.

After about three months I had collected the available information and written it up for publication. One sticky point centered on how to treat the blunders that Dobzhansky had made. We finally agreed that we would not dwell on the point, but would simply remark that his observations were in error. I simply said: "The observations of Dobzhansky and Socolov (1936) on azteca failed to include the longer salivary gland chromosome arm of their 'C' chromosome, a circumstance that invalidates their conclusion that 'some genes found in D. azteca in a single linkage group must belong to different ones in D. pseudoobscura; conversely, some genes linked in D. pseudoobscura may be expected to be independent in D. azteca.'"

Sturtevant wrote the final two sections of the paper and it went off to press. He put my name on it as second author even though my contribution was only that of a paid writer, having made no intellectual contribution whatsoever (Sturtevant and Novitski 1941b). At the time I could not have guessed that a most convincing demonstration of the integrity of the chromosome elements would eventually come from a comparison of the human chromosome complement with that of the chimpanzee, where there are many inversion differences, but, except for one unusual end-to-end telomeric fusion, no translocations (Appendix E).

SECTION 5

STURTEVANT AND DOBZHANSKY AT ODDS

The Rift Deepens and Dobzhansky Departs

There is no question that by this time (1939) Sturtevant had been convinced that the Drosophila group would be better off without Dobzhansky. It is not clear what consideration, or combination of considerations, may have predominated in Sturtevant's decision. Among other things, it may be that he had become aware of Dobzhansky's religiosity, either previously hidden or just developing, which Sturtevant, as an atheist, felt was incompatible with the fundamental philosophy of a true scientist. He may have felt uncomfortable with Dobzhansky's acceptance as guest investigators individuals of whom he did not approve (such as Pio Koller, who had a religious background).

Some historians have suggested that he may have felt envious of Dobzhansky's increasing stature in the field, resulting in large part from the wide acclaim given his book *Genetics and the Origin of Species* and that this threatened his position as leader of the Drosophila group. There is no information available on Sturtevant's reaction to Dobzhansky's book. It seems likely that with his greater exposure to the developments in the field, from history, from the published literature and from personal experience, he would have some differences of opinion with Dobzhansky's interpretations. For instance, one case that must have been obvious to Sturtevant comes in Dobzhansky's discussion of the two races of D. pseudoobscura where he states that their ranges "overlap without the formation of hybrids in nature" (page 316). Since it was known from the original description of the two forms by Lancefield in 1925 that they can intermate and produce hybrid offspring in the laboratory, it seems unlikely that this statement could be true and, in any case, seems to imply an omniscience that few workers can claim. Any student making such a statement in a term paper would find, upon its return from the instructor, that statement redlined with a comment in the margin: "How do you know?" This simple misstatement does not fall into the category of the unfortunate trivial error every science writer will inevitably make. It is an important conclusion on a matter about which the two would have strong differences of opinion, as we shall see later. There are the more scientific differences. With the progression of time, it must have become more and more obvious to Sturtevant that Dobzhansky was more naturalistically inclined and was less interested in the kind of experimental genetics research that had provided the fundamental information about the nature of the genetic mechanism and that had made the Morgan group renowned worldwide. Sturtevant might have detected a shift in Dobzhansky's interest from that of an experimentalist to that of an expositor, a point to be considered in detail by the contribution of Prof. R. C. Lewontin reproduced later in this book. Finally, Dobzhansky's dynamism and enthusiasm included an interest in forming a small self-contained empire independent of the other members of the Drosophila group, headed off in a direction that Sturtevant would have considered to be unacceptable.

However it would not be possible to suggest to Dobzhansky directly that his departure would be desirable. Sturtevant was the sort of person who would avoid any direct personal confrontation at all costs, and even if he did, he might conceivably have been overruled by colleagues, particularly T. H. Morgan (13-1), the administration, or academic protocol. The wide acclaim given to Dobzhansky's book had elevated his stature considerably. It was necessary for Dobzhansky to leave of his own volition. There were several causes, perhaps each by itself trivial but jointly of some significance, that led to the incompatibility, but it is possible that Sturtevant, who now realized that Dobzhansky had made serious errors in his repeated claims that closely related species differed drastically in their chromosomal makeup, had in a Machiavellian twist chosen me, as a former student of Dobzhansky, to add to his discomfort.

Why these inaccurate observations on the prevalence of gross chromosome differences in closely related species were open to serious criticism needs some explanation. That they were mistaken might be, of course, a matter of some slight personal embarrassment to him, but in itself is not that all that important. That they were immediately published without any effort to check on what were clearly unusual phenomena becomes more problematic. Then Dobzhansky's failure to have either of his students (Helfer or me) follow up these observations when he assigned us problems becomes questionable.

But, in my opinion, the most curious aspect of the situation is that Dobzhansky seemed to realize that according to the theory of population mathematics this sort of event could happen only very rarely, and that if no extraneous cause could be found, it would mean that the mathematics of populations worked out by such persons Sewall Wright, J. B. S. Haldane, and R. A. Fisher would be open to question. The explanation for this may rest in the differentiation of the intelligences in Dobzhansky, his excellence in the areas of naturalistic and linguistic, accompanied by a concomitant depression in the logical-mathematical.

Departure of Dobzhansky from Caltech

Dobzhansky left Pasadena without fanfare. Sturtevant was away at the time and as far as I know there were no goodbyes to or from any of the students. I noticed his absence but thought that he had gone on a trip somewhere. It was a few months later that I learned that he had left permanently for Columbia University.

The move from California to Columbia viewed from the outside would undoubtedly appear as an improvement in Dobzhansky's status, his income, etc., as all such moves are, and would be accepted without comment or question. In this case, however, there were definite negative aspects. Dobzhansky had come to love the variety of the California landscape, the mountains, the deserts and all in between. As he says on p.406 of his Oral History: "Death Valley region is an extremely beautiful region, so here was satisfying one's wanderlust, while collecting material for study. That was an extremely fascinating occupation, traveling in this arid desert, climbing the mountains . . ." He gloried in the outdoor life and used collecting trips as a legitimate excuse to spend as much time as possible in the outdoors. And the collecting trips supplied him with the material he needed for his extensive laboratory work and publications.

The canyons of the Manhattan skyscrapers were a poor substitute for the canyons of the Western mountains. Add to this the trauma involved in moving the family, his daughter Sophie, and his wife Natasha (and her mother) from the casual lifestyle, sunny climate, orange orchards, and pure air of the Pasadena of 1940 to the rough and tumble life of New York City. There are of course many compensations, as any native New Yorker would be quick to point out, and Dobzhansky lists some of these in his account.

Nevertheless, there obviously must have been compelling reasons for this move: Dobzhansky gives two at different points in his Oral History Memoir, neither convincing.

The reason he gives first is that Morgan, with whom he had come to work back in 1927, was no longer working with Drosophila and so the association no longer had the strong attraction that it did initially. This however would not appear to be a compelling reason to leave. Morgan had not really been an active worker with Drosophila after coming to California and Dobzhansky himself had been severely criticized by Sturtevant for making some off hand derogatory remarks about Morgan.

The second reason that he gives for the move is that he thought that his friend Pio Koller was treated in a hostile way by Sturtevant. It is probably correct that Sturtevant was unfriendly towards Koller. Sturtevant was opposed to the thought that anyone with a truly scientific mind could at the same time be devoutly religious (but he must have made an exception in the case of Gregor Mendel!). But this also does not hold water as an important reason for Dobzhansky's departure. I had been with Koller and Dobzhansky at Cold Spring Harbor as Koller was returning to Europe in 1938, and several weeks later, as we were traveling to Pasadena by train, Dobzhansky was clearly contemplating an indefinite stay at Caltech. This included taking on a new graduate student (me) and setting up a research project that would necessarily take several years. Since I had applied to become a student in the department initially, and not a student of any particular person, this acquisition of me as a graduate student by Dobzhansky certainly suggests not only an expectation of indefinite tenure, but also implies a feeling of security in his position. It would have been unconscionable for any faculty member to acquire a beginning graduate student when there was any expectation of leaving, without informing that student of the uncertain future.

It is my conviction that he left because he no longer felt welcome, but not for the reasons he gives. The rather pointed criticisms by those of us working in fields closely related to his recent papers were unpleasant, but the fact that it was Sturtevant who suggested that we work in those areas probably conveyed a message that he could not ignore. It would have been impossible for Dobzhansky not to have felt uncomfortable at the repeated assaults on his work by those of Sturtevant's students working on population problems and not to have realized that this activity, if not actually being promoted by Sturtevant, at least was being done with Sturtevant's acquiescence. Under those conditions it must have been clear to him that there existed a problem of compatibility.

That he was embarking on a population genetics program of which Sturtevant expressly disapproved could have added appreciably to his discomfort. Sturtevant's effective nullification of his decision to terminate a graduate student, with its implication of a hierarchical structure, may have conflicted with his view of his own authority and the dignity of his professorial rank.

Further, the natural desire to expand his population work faced a bleak future in Pasadena. A move would bring brighter prospects for increased funds and, more importantly, more assistants, particularly bright graduate students. These would have added to his decision.

In retrospect, the move to Columbia away from the dominating influence of Sturtevant was probably the best thing that could have happened to Dobzhansky. He was now a free agent and as such could develop his area of interest without restraint.

During the period immediately before his departure, neither I nor any of the other graduate students were aware of the differences between the two. At no time did the students hear either speak ill of the other. I myself was so innocent of any friction that I did not hesitate, in 1943, to visit Dobzhansky in his lab at Columbia University, and only then did I realize, by body language more than anything else, that I was an unwelcome guest, and I terminated our visit earlier than I had planned.

That there were substantial differences between the two, and that Dobzhansky was smarting from his last years at Caltech, however, became most evident several years after he left for Columbia.

A Fly By Any Other Name

The first (as far as I know) indication of outwardly expressed hostility between the two came in 1944, four years after he left. It centered on Dobzhanky's naming Race B of Drosophila pseudoobscura as Drosophila persimilis.

The two races of Drosophila pseudoobscura were first identified by Donald Lancefield, who had been a student at Columbia during the period of the 20s, and who was a personal friend of Sturtevant. Lancefield discovered two forms of Drosophila obscura (as pseudoobscura was then known) in some collections from Oregon in the early 1920s; these he named Race A and Race B of obscura, the name obscura being changed to pseudoobscura by a Russian worker in 1929. Lancefield's first paper in 1925 was entitled "An interracial cross in Drosophila obscura producing partially fertile hybrids" and this was followed by a more extensive analysis in 1929. It was undoubtedly Sturtevant who persuaded Lancefield that the two forms should not be considered different species, since they were morphologically indistinguishable, mated with each other, and produced some fertile female hybrid offspring, thus making it possible for genes to be transferred from one of the races to the other.

Dobzhansky has a somewhat different view. He describes a species as that stage in the evolutionary process "at which the once actually or potentially interbreeding array of forms becomes segregated in two or more separate arrays which are physiologically incapable of interbreeding." But then later states that "the races are not distinguishable morphologically, this being the only reason why they are not classified as distinct species." (Dobzhansky 1937). He also states that two forms, to be considered different species, must be completely intersterile.

Apparently the complete sterility of the male progeny of crosses of the two forms is being considered adequate qualification for the description of "complete" sterility. However, fertility, even partial fertility, of the hybrid females would allow for the transfer of genetic material from one form to the other after mating with males of either of the two races.

Possibly the impression of complete separation of the two forms is supported by the observation that flies collected in the wild show no evidence of hybridization which would be evidenced by the production of progeny with the characteristic chromosome types, different for the two forms. However, it should be noted that the procedure for collecting flies may very well have precluded such an observation. My experience, which followed standard practice at the time (1939), was to remove all flies from a trap, discard the males and sequester the females individually in separate vials, whereupon their progeny would form the basis for a new and distinct line. If a female of one race had mated with a male of the other, all the male progeny resulting would be sterile and the fertile or partly fertile female progeny could produce no viable eggs. Such a culture would be simply discarded as sterile, and of no interest (For a more up-to-date discussion see endnote14-1).

With respect to the renaming of Race B, it should be emphasized that Lancefield was a good personal friend of Sturtevant's, not just from their Columbia days together but later by their joint ownership of a dwelling at Wood's Hole, which they shared by occupying it each on alternate summers. This close relationship was well known to Dobzhansky.

So, a few years after arriving at Columbia, Dobzhansky and Epling (1944) renamed Race B as "Drosophila persimilis," something he could not have done while he was at Caltech. From one point of view, it must have been a demonstration of his newfound scientific freedom, but it was more than that.

It is commonplace for systematists to name new species after individuals, preferably someone who was involved in the discovery, or working in the field, etc. (14-2). The name Drosophila lancefieldi was in fact given to Race B by a worker named Ginsburg but, according to Sturtevant, Dobzhansky and Epling avoided Ginsburg's intention and failed to recognize that Ginsburg in fact did give "an unequivocal bibliographic reference," an essential step in naming a new species.

Although Sturtevant objected in principle to Race B being named as a species, to name it Drosophila persimilis and not Drosophila lancefieldi was a personal affront that he could not overlook.

Sturtevant immediately responded to this obvious offense with a paper in Ecology (1944, reprinted at the end of this book) in which he took Dobzhansky to task for making "a clear misinterpretation of Ginsburg's intention" and for using questionable statistical methods, such as applying the equivalent of fifth power of wing measurements to try to bolster his claim that the two forms were morphologically different.

Sturtevant did not mention the other consideration that must have played a role in the initial decision to consider these two forms as races rather than as distinct species, that is, that they produce partially fertile female hybrid offspring, thus allowing for the transfer of genes from one group to the other. Dobzhansky himself gives as one of the essential criteria for designating two forms as separate species the *complete* sterility of any hybrid offspring, which is not the case for the two races of D. pseudoobscura.

Thus, these pronouncements, made in Dobzhansky's classic work of 1937, might very well have contributed to Sturtevant's growing conviction that the two had different and irreconcilable points of view.

In the same issue of the journal Ecology Sturtevant avails himself of the opportunity to criticize another paper by Dobzhansky and Epling, one involving a study of chromosome distributions in natural populations of D. pseudoobscura. In his review, Sturtevant points out that their results, which they interpreted as the action of natural selection on chromosomes of different genetic compositions, could be just as well explained by the action of the strong winds well known in the area of observation, and he suggests ways in which their observations might be improved.

Never before had Dobzhansky's work been exposed to such biting criticism from a well-established worker in the field. That Dobzhansky was disturbed by this intrusion into the usual smooth acceptance of his work became obvious from the critical treatment he later gave Sturtevant in his Oral History Memoir. To him, this was a personal attack which would not be forgotten.

Dobzhansky's "Oral History Memoir"

When I became aware of Dobzhansky's Oral History Memoir, dictated in 1962 and preserved on microfiche, I was curious to see how he would handle those incidents that must have been personally distressing to him that occurred prior to his departure from California.

What I found was that there was not a mention of them. In fact, the entire period of four years (from 1936 to 1940) was almost entirely obliterated from consideration. Not only events, but all persons in Pasadena with whom he had come into contact, are cut.

His Oral History Memoir was recorded by Ms. Land, who occasionally interrupted his dictation for further details or clarification.

When he mentions the field trips during this period, for instance:

Question from Ms. Land: "Were you alone on this trip?"

"Getting flies. No, not this trip, it was definitely 'trips' plural. It took just about two years to collect this material. As a matter of fact, by that time our little daughter [15-1] . . . And of course there were always good companions. There were several graduate students in Pasadena who went, partly for the fun of it, partly for the sake of learning the collecting [Figure 7]. Then, finally, there were several foreign fellows, coming to the laboratory for winter season, and I think that about that it would be better to talk next week."

The scarcity of information about those collecting trips was in stark contrast to the detail given earlier about other activities must have been obvious to Ms. Land, who then pressed him for more details. He moves on to mention his daughter on collecting trips, vaguely refers to graduate fellows and foreign fellows and quickly terminates the session. Although he says that he would get back to the subject in the next session, in fact he never does.

Figure 7. At the top of Telescope Peak in the Panamint Mt. Range near Death Valley, 10 June 1939. Survivors of the seven mile hike include E. Novitski, Esther Ellen Rudkin, B. Stewart, Th. Dobzhansky and A. Socolov.

The name of the eminent German cytologist Hans Bauer does not appear, even though Dobzhansky collaborated with him on an examination of the chromosomes of hybrids between D. azteca and D. athabasca and with Bauer as senior author published a short abstract which indicated, but did not mention, the error of the previous observations of Dobzhansky and Socolov. There is no mention of Waddington, visiting with his wife from England, with whom he and the Rudkins (and I) spent a weekend in the San Jacinto mountains, or any of the other visitors to the department during that time. Charles Rudkin, his daughter Esther Ellen and his son George, who worked with him as an undergraduate on GNP I and II, are cut. Students working in the lab, like Bob MacKnight and Dwight Miller, are cut. Students for whom he had some responsibility in their Ph.D. work: Klaus Mampell, Bill Hovanitz and of course E. Novitski all are cut. And this was the most unkindest cut of all: Bob Helfer, his graduate student who took care of him at the time he broke his leg (with great skill, Dobzhansky had told me during the train trip in 1938), is not mentioned at all, although the accident itself is described in considerable detail (Oral History Memoir pp. 394-395).

One can only wonder at the reasons for this astonishing amnesia. I myself believe that he felt so strongly that he was mistreated by those of us working on population problems under Sturtevant (and, therefore, by Sturtevant himself) that he wished to avoid the trauma of recounting the entire situation and decided to try to sidestep any discussion by simply blanking the whole period.

Also, it would have been perfectly understandable for him to avoid any unnecessary public acknowledgement of the errors he had made in his cytological work during that period of time. If so, he succeeded admirably. The controversial work is buried in technical papers that are quite complex and only a few dozen or so of us who were intimately involved in this research would understand the nature of the difficulties; most casual readers of the genetic papers would never be aware that there was any problem at all.

In his descriptions of his former colleagues, he would appear to be unusually harsh. It must be kept in mind, however, that he was no longer favorably inclined towards Sturtevant after the latter's critical papers in Ecology and that he was unsympathetic to the kind of experimental research in which Sturtevant excelled, preferring a more naturalistic flavor for his chosen direction. His hostility towards Sturtevant bubbles to the surface; there appear in his account several comments about the latter days in the lab and Sturtevant does not fare well.

On p. 270: "both Bridges and Sturtevant were merely assistants in the Carnegie Institution. Neither Bridges nor Sturtevant had an education. Neither Bridges nor Sturtevant were people of culture beyond Drosophila genetics, but they were supreme geneticists. They were supreme specialists. As I said, Morgan was something very different. He was a man of highest intellectual standing."

On p. 272: "Morgan clearly had to present these discoveries in books. The others could not. That they could not is attested by the fact that they never published any books."

About Sturtevant Dobzhansky says: "he is not the kind of person who can write an obituary or anything other than a scientific paper." (Oral History Memoir p.251)

At one point, after commenting on the genius of Bridges, he says (p.268): "Sturtevant . . . still living . . . was a person about as different from Bridges as you can imagine. Very matter of fact, very self-possessed, a steady worker, no flashes, no sparks, just good steady work every day, including Christmas Day, New Years Day, always in the laboratory, always counting flies."

At that moment in the dictation Dobzhansky seems to have forgotten that it was Sturtevant's flash of inspiration about the relationship of overlapping inversions in constructing a phylogenetic tree that started the work on the inversions in Drosophila pseudoobscura, the development of the details of which was one of Dobzhansky's great achievements.

(About Morgan he later says on p. 276: "He was in the laboratory every day . . . I mean every day, including Sunday, Christmas Day, New Year's Day . . .")

One might wonder how Dobzhansky knew that Sturtevant and Morgan were at the lab every day. What the average reader would not appreciate is that all those who did lab work with Drosophila, especially melanogaster with its short life cycle, found it necessary to go to the lab every day without exception to do the daily (and sometimes twice daily) collecting of freshly hatched females before they could mate with some male of unknown genetic constitution. Anyone who found this routine unacceptable could easily move to some other species with a less demanding life cycle or to some other branch of genetics. After a dozen or more years of this routine this becomes so much part of the biological clock of fly geneticists that they, like Pavlov's dogs, automatically go back to the lab every day even when it is not necessary. And of course they could not skip weekends or Christmas or New Year's Day unless they were out of town or in the hospital!

Also, the description of this lab routine as "work" is highly misleading. If a person did not consider this routine to be a necessary part of an activity that was pure pleasure and fun he would get out of the field pretty fast.

I have a distinct recollection of sitting in Dobzhansky's lab on many Sunday mornings, when we were not spending the weekend in the mountains on a field trip, trying (unsuccessfully) to complete the Los Angeles Times crossword puzzle while he carried on a conversation as he whipped through slides of pseudoobscura chromosomes, a job which did not have to be done on a Sunday. Of course when his "Oral History Memoir" were recorded, some 26 years had passed and perhaps his routine had changed and his memory (which seems to have failed him in other ways as well) was clouded.

When he finished dictating these memoirs in 1964 he stipulated that no one should be allowed access to them without written permission. This was a curious prohibition and suggests a preference that certain persons not see them for some time. In fact, he states this explicitly when he makes a derogatory comment about a worker, one he describes as a "stupid person," at the nuclear facility at Oak Ridge, Tennessee. However, since Sturtevant is directly denigrated a number of times, one can reasonably conclude that Sturtevant is the principal person in question. This prohibition was lifted in 1972, two years after Sturtevant died.

Research and Writing Style of
Th. Dobzhansky

After moving to Columbia University in 1940, Dobzhansky could now, as the prime focus of the growing field of evolutionary biology, feel free to write for an audience of population geneticists, naturalists, and humanists, and to philosophize in ways that would have been unacceptable to Sturtevant.

The series on the study of wild populations of D. pseudoobscura would be expanded considerably, involving several full-time assistants and graduate students, and culminating in more than forty papers. Having a predisposition to be critical of this kind of work from my earlier experiences as a graduate student with a similar project for a thesis, and because my time was monopolized by other genetics interests, I did not read subsequent publications in this area myself and so, being unable to evaluate these works (with one exception), I must leave this to others.

Most of the descriptions of Dobzhansky's works are highly laudatory, and their authors are often nonbiologists (anthropologists, historians of science) or, when biologists, belong to the category of the linguistically rather than logical-mathematical gifted. An exception is Prof. Richard Lewontin, a student of Dobzhansky, who happens to be one of those rare fortunate individuals who is endowed both verbally and mathematically. He is an active worker in the field of population genetics, having received his Ph. D. degree with Dobzhansky. He admires his mentor's work, and his person, and can hardly be considered negatively biased or in any way predisposed to treat him harshly. He has, among many other accomplishments, studied the papers in Dobzhansky's Genetics of Natural Populations series and written a critique of each (Lewontin, Moore, Provine and Wallace 1981).

So it may come as a surprise to learn that his evaluation of Dobzhansky's procedures is that they are not scientific experiments in the commonly accepted sense, but are rather demonstrations of the relevance of population studies to evolutionary theory. Lewontin's perceptive analysis (Levine 1995) is so revealing and pertinent to the discussion in this book that it is reproduced, with permission, as a special section at the end of this volume. Because the discussion becomes highly technical in parts and probably would be unintelligible (or uninteresting) to the nonbiologist, a brief summary is included here, along with some pertinent quotes from his account.

"Theodosius Dobzhansky is thought of as one of the founders of experimental population genetics and especially as the person who brought together laboratory experimentation and observations of the genetics and behavior of natural populations in an attempt to explain the dynamics of genetic variation in nature . . . Dobzhansky is always placed at the experimental end. I argue in this paper for a very different view of Dobzhansky. I claim that he was in fact a theoretician whose entire intellectual program was theoretical and conceptual but that, lacking the abstract mathematical tools that are the stock-in-trade of the conventional theoretician, he used the only tool at his disposal, the manipulation of living organisms."

About his inadequacies in the mathematical intelligence area, something we have alluded to several times previously, Lewontin says: "Dobzhansky himself was innumerate, at least he appeared to be. He could not carry out and claimed not to understand the simplest algebraic manipulations and was unable to use the tables of the standardized normal curve to draw a normal deviate to draw a normal curve with a known mean and standard deviate. He would on occasion demand from a theoretical colleague a formula that could be put to some use in manipulating data, but he never asked how the formula was arrived at. This picture of Dobzhansky, familiar to his students and postdoctoral fellows, comes as a surprise to those who know his work only from his publications." But, "It is not surprising that the body of Dobzhansky's work on population genetics would give the reader the impression of someone very familiar with and at home with the most advanced theoretical biology of the time. A closer examination of the papers and knowledge of Dobzhansky's actual practice reveal the actual situation. Dobzhansky collaborated constantly with theoreticians."

"However, Dobzhansky never depended upon theoreticians for the conceptualization of the experiments and never allowed them to influence his interpretation of the results. This last point is material to the claims that Dobzhansky was a "theoretician without tools." Dobzhansky knew at all times what the experiments were intended to demonstrate and what the conclusion was from the results. Indeed, if my claim is correct that the experiments were largely *demonstrations* rather than explorations, these conclusions were already in existence *before* the experiments were done. The role of the theoretician was to use his expertise to draw the rigorous quantitative connection between the data and the conclusions that Dobzhansky had already decided upon. Any real independence on the part of the theoretician to draw other conclusions from the data was not tolerated, and this applied even to Sewall Wright, whom Dobzhansky believed to be "one of the two greatest living geneticists."

The reader interested in the complete argument presented by Lewontin is urged to read his detailed exposition reprinted in the latter section of his book. For the moment, we are better able to understand the workings of Dobzhansky's mind-set. In particular, we are in a better position to appreciate his attitude towards taking liberties in presenting data in publications, a point that will be taken up in the next chapter.

Population Cage Experiments

A good example illustrating Dobzhansky's view of "experimental" results is given by some work with population cages.

A population cage, originally devised by L'Héritier and Teissier (1934), is simply a large box into which many hundreds or thousands of flies are enclosed with a good supply of food and allowed to breed freely over a long period of time.

Usually there are two identifiable types of flies introduced in arbitrary proportions initially, with samples being withdrawn at intervals, and their frequencies observed to see if the proportions change over time. When they do change, this is usually interpreted as evidence that the one doing better has a more "fit" genetic complement under the conditions of the population cage at the time.

Although Dobzhansky had, in 1939, dissuaded Dwight Miller from such an undertaking using different chromosome types in Drosophila algonquin, by the 1950s he apparently had a change of heart (17-1) and embarked on a series of such observations himself, using wild caught strains of D. pseudoobscura which differed in the genetic compositions and structures of the chromosomes they carried.

I myself had no interest in the papers that resulted and paid no attention whatsoever to them. However, several years ago while leafing through a book looking for a reference I came upon a drawing showing his results from such cage observations (Dobzhansky and Pavlovsky 1953). My attention was drawn to the neatness of the experimental results. A graph showed a continuous change in four independent cages, all started with a low frequency of one of the chromosome types and proceeding in lock step to a very high frequency of that type. The similarity of the four sets of observations over the course of a year was a cause for wonder (Figure 8). It was the sort of graph that one might see submitted by a bright but lazy undergraduate in a college course. In most cases when an investigator attempts such a long range experiment lasting a year with several, in this case four, replicates, one can reasonably expect that a deviation of some sort in at least one of the sets will mar the complete uniformity of results. It occurred to me that perhaps he was working, not with four independent population cages, but, after some commingling, with effectively fewer, and I developed some thoughts on how this might have legitimately come about.

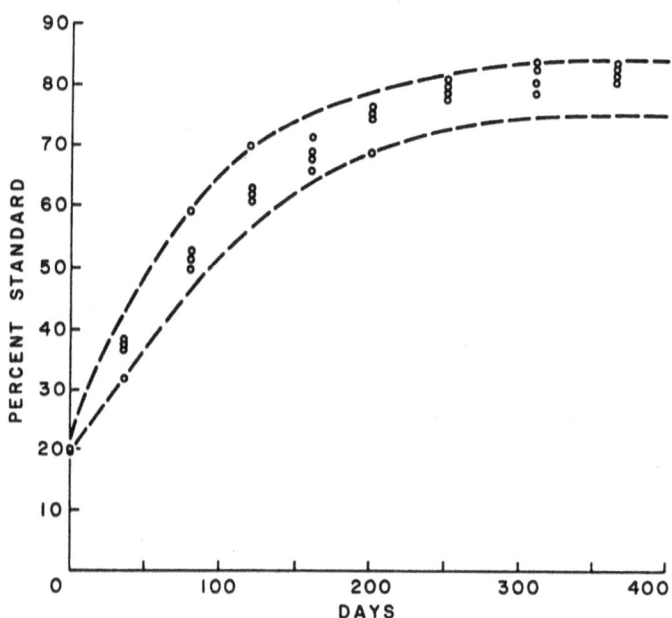

GNP SERIES XII

Figure 8. Reproduction of a graph from Dobzhansky and Pavlovsky (1953) showing the changes in the frequencies of chromosomes carrying different chromosome sequences. Samples from four population cages were followed over the course of a year. The unusual consistency of the observations led to a question as to whether some additional factor was at play in these observations, and comments from some workers present during that time indicate this to be the case.

Since I knew personally some of the students who worked in the Dobzhansky lab at Columbia at about that time: Colin Pittendrigh, Stanley Zimmering, George Streisinger (17-2), Dick Lewontin and a few others, I wrote a number of letters with the simple question: Could they provide me with information about the conduct of the experiments which might throw some light on the uniformity of his results? One of the replies was quite illuminating.

Lewontin wrote: "in addition to the two inversion types that Dobzhansky was following in the cage there was a third type in a rather high frequency that Dobzhansky had never noticed." This came as a surprise, for the description of the experiment gave no indication that one of the lines had experienced a disaster. The appearance of an extraneous chromosome configuration in any one of the four lines would require the destruction of that line, because there would be no way of identifying flies with the offending chromosome type without cytological examination; its presence would prevent any count of the two competing types and would preclude any conclusions that might be drawn from those counts. Even though the graphs, data, text and statistical analysis indicate that the four lines were independent and continuous, the contaminated line must have been replaced with specimens from one of the other three lines (17-3).

Lewontin agrees that the contaminated cage must have been discarded, that "a new cage was made up from one of the others or perhaps, from a mixture of the others" and that, after careful examination of the sequence of events during the course of that set of experiments, he pinpoints the time of the discovery of the extraneous chromosome and the consequent commingling, at July of 1952.

"the real issue is not whether the cages were sampled to make up the new cage, but what was to be made of the chromosome frequencies shown in [their] figure when in fact there was another chromosome segregating."

The source of the numbers that appear in the paper as representing the figures at the time of commingling is open to speculation.

Thus my original suspicion of commingling of lines was confirmed. This contamination, however, brings up another question. How did the contamination originate? The accidental penetration of a population cage by extraneous flies seems out of the question. Not only are the cages necessarily constructed securely to prevent the movement of flies in and out, but if one or two foreign Drosophila got in, how could their characteristic chromosome configurations reach a high frequency quickly in that population cage? One could imagine that they were introduced in quantity by a disaffected co-worker, or by a malicious student. This would represent a unique event; in my four decades or more of laboratory work with Drosophila, I have never heard of such sabotage, either in my lab or in others. The only reasonable explanation is that it was an honest mistake, probably by an inexperienced laboratory assistant. It would be understandable from the following considerations.

Before I became aware that commingling in fact did occur, but was searching for an explanation for the apparent closeness of the four presumably independent experiments, I searched in my mind for a possible explanation and it dawned on me that there was in fact a simple one. One word:

Mites!

Mites are the scourge of every Drosophila lab. Every instructor who has taught an elementary laboratory course in genetics involving Drosophila has had the experience of finding an old culture bottle, abandoned by a student (usually over Christmas vacation), now crawling with the beasts. They are persistent and ubiquitous. They have a migratory stage at which they attach themselves by the dozens to the adult fly making it impossible for it to fly, reproduce, etc. They appear from nowhere and many Drosophilists are convinced (incorrectly, I believe) that they arise by spontaneous generation. When they do, they can be eliminated by the rapid transfer of adult flies to successive culture bottles; the mites will leave the fly for the better culture conditions in the new cultures. This is easily done with Drosophila melanogaster because the life cycle of the mite is slightly longer than that of D. melanogaster. It is more difficult with other species like D. pseudoobscura which have a longer life cycle. I have had miserable experiences with mites infecting stocks of both species.

However, once they proliferate in a population cage, (easily done because it is readily penetrated by the tiny mites,) they become a real threat. When they get out of hand, the cage must be sequestered and autoclaved (or, alternatively, heat treated) until all signs of life are extinguished. Otherwise, it becomes a potent source of more mite contamination.

Not only would mites be present naturally in the woodwork of Schermerhorn Hall and in the lunch boxes of students (mites are particularly fond of cheese sandwiches), they would also be introduced continually when collections were made in the wild, where they would infect perhaps one fly in twenty. Many of the collections were made by friends and colleagues in the West and sent by post to Columbia University. Dobzhansky tried to control the mites chemically, which might have worked to some extent. But he must have contaminated Schermerhorn Hall at Columbia University to the hilt. It is inconceivable that they could have run four population cages for a year without even once running into a serious mite contamination inside the cages originating from outside.

Lewontin states, in a letter on this general subject: "There certainly were mites in the culture room at Schermerhorn. I am not sure I would talk about "outbreaks" so much as a constant population of mites. They got particularly bad in a particular culture when it was neglected, but they were there all the time and in reasonable numbers. Population cages especially lead to mite populations."

They undoubtedly had a way of handling this serious problem routinely. Olga Pavlovsky, the co-author of the paper and one person who would have known precisely what measures were taken when mites became a serious problem, unfortunately died in 1985, long before this became an issue. One solution involves simply refreshing the contaminated cage with flies from one of the other good cages, or, if the infestation is completely out of control, to destroy the offending cage and replace it with a new cage with flies from one of the other cages. In fact, this can happen several times, and each time a contaminated cage is refreshed with flies from an uncontaminated cage, the variation that one might intuitively expect from four independent experiments is reduced.

This conclusion also provides a possible explanation of how the extraneous chromosome got into one of the cages, and when detected, had reached a high frequency. We assume that in refreshing or replacing one of the contaminated cages, one of the workers inadvertently used the wrong source, and put in flies with the wrong chromosome accidentally.

Anyway, my conclusion is that there were not four independent populations studied over a period of a year. This is confirmed by the discovery of an extraneous chromosome which necessitated the destruction of one of the lines and its replacement by subculturing one of the other three, and it is my conclusion that there were some (number indeterminate) cases of mite contamination and subsequent commingling which brought the points on the graph closer together.

About the inconsistency of the known events during that experiment with the published results, Lewontin says:

"At any rate, despite the finagling, which as I say was not untypical of Dobzhansky, the data, as a whole, certainly do support his claim. Nothing would be lost if one simply left out of all of the data one or the other cage. Any one of the four cages could be simply be removed from the data and nothing would change. I do not mean to excuse the sloppiness and even vaguely dishonest behavior of Dobzhansky in this respect, but only to say that he had rather lax standards when the truth appeared to him to be self-evident. We would not want to encourage people, in general, to behave in this way. In the present case, at least, no damage to the truth was done."

Then the question arises as to whether Dobzhansky was aware of the consequences of this subculturing. If there is one aspect of Dobzhansky's intelligences about which there appears to be some agreement, it is that he was deficient in the logical-mathematical skills (this apparently being crowded out by a superb capability in the linguistic and naturalistic areas.) In the first place, he might have had no reason to believe that taking a sample from another population cage, a parallel one, to replace or augment a contaminated one, would not be a completely legitimate operation. Second, if he was at all uncomfortable with this maneuver, he might reason that external reasons justified it and, in any case, it was such a minor matter that no one would notice and if anyone did, they would not be so petty as to make an issue of it.

Is this apparent discrepancy important? I myself don't think so. No new phenomenon, principle, or great conclusion rests on the similarity of the four runs. The only difference is that the conclusion that these experiments are completely repeatable would lose some of its weight.

Others more intellectually endowed, or with preconceived ideas of the nature of the scientific enterprise, might take a less forgiving view. For instance, Feynman (1999) gives his opinion in a section devoted to nonrigorous investigations entitled "Science which is not a Science": "You see, I have the advantage of having found out how hard it is to get to really know something, how careful you have to be about checking the experiments, how easy it is to make mistakes and fool yourself.

I know what it means to know something, and therefore I see how they get their information and I can't believe that they know it, they haven't done the work necessary, haven't done the checks necessary, haven't done the care necessary. I have a great suspicion that they don't know, that this stuff is [wrong] and they're intimidating people. I think so. I don't know the world very well, but that's what I think."

SECTION 6

THE PLEASURE OF
FINDING THINGS OUT

The Joy of Figuring Things Out

During the course of a career in science every investigator has experienced the "pleasure of finding things out" (Feynman 1999). This is generally considered to be one of the great attractions for young people contemplating such a career, and one of the substantial benefits accruing to the advanced practitioner.

As a subset of this category, there is another aspect, which I have used to designate this chapter. It comes from those rare cases when it is possible to take advantage of the intellect, more often than not by chance, to reach a conclusion not easily arrived at during the routine of biological investigation. This is not to imply that most scientific enterprise does not involve some use of the intellect, or that this occasional flash of insight somehow transcends ordinary thought processes. Nor does it deny that most worthwhile scientific research, from the standpoint of the good benefiting society as a whole, centers on the patient routine collection of data, or the steadfast adherence to a program of investigation carried out over a long period of time. It simply refers to those relatively rare instances where, in an unusual flash of intuition, several neurons usually separated in the vast complexity of the brain come together and suddenly produce a connection with consequences far greater than the sum of the two individually. This is popularized in today's cartoons by showing a lamp bulb suddenly lighting above the recipient's head.

The reaction of the individual is one of pure joy. Classically the story is told of Archimedes, who faced the problem of determining whether the newly fabricated golden crown of the King of Syracuse contained an adulterant, silver. It is said that the solution occurred to him during his bath and that he was so overcome by the realization of his intellectual achievement that he ran down the street in a state of undress, shouting, "Eureka!"

The shout of Eureka is not uncommon in laboratories these days, although usually in the solution of problems far more mundane than that of determining the amount of silver adulteration of a golden crown. But when it is heard, it is usually an indication that the intellect has had some role in the solution of a problem in an otherwise humdrum routine enterprise. And sometimes it is never heard, for the constraints of biological investigation and the limitations imposed by the necessity to conform to the dictates of the nature of the living organism seem to preclude any flights of the imagination. Indeed, some might argue that studies of biological phenomena are hampered by the intrusion of the intellect (see, for instance, the views of J. C. Lucchesi and W. J. Peacock in Lucchesi 1994) and that the science of biology should be confined to matter of fact observations in those areas that are considered to be "on the cutting edge." More intelligent and enlightened views, held by G. H. Hardy and C. Stern, are discussed in Chapter Twenty Six. In particular, this aspect of personal fulfillment may be incomprehensible to that group of otherwise competent writers known as "historians of science" who would never have had the opportunity to experience the personal satisfaction that comes from such a highly personal, even narcissistic, achievement, and who are therefore reduced by necessity to evaluating research accomplishments of others by counting the number of their publications.

The writer has had the fortune to be involved to some extent in an area where some amount of intellectual activity was possible and, to a greater or lesser degree has experienced "the joy of figuring things out." With it of course goes the occasional failure but these are soon forgotten, much as the high school athlete will savor the memory of his having broken the school record for the high jump, or the 100-yard dash, and forget the many times when in fact he tried and failed, or that his school record pales in comparison with regional or world records. The rest of this section deals with some of those moderate successes.

Muller's Euphoric Joy

Muller paced back and forth across the small living room floor. He was concentrating on finding an answer to a genetic riddle involving Drosophila which I had posed as a challenge. At intervals of a few minutes he made suggestions for possible answers, all of which he knew were inadequate. When finally he gave up and I explained the solution, he exploded in what can only be described as a fit of ecstasy. His pacing changed to a childlike skipping, with his arms overhead, clapping his palms together. This was my introduction to one of the most remarkable geneticists of the early twentieth century. His concentration and enthusiasm were clearly evident, and I immediately understood how such a personality would be a thorn in the side of any co-worker who might have to tolerate him over a period of time.

Some historical background for this account follows:

While at the University of Missouri after the war, I worked with Allen Griffen, a recent graduate of the University of Texas, who was starting a program analyzing X-ray induced rearrangements in Drosophila. While from one point of view this program had some merits, it was greatly hampered by the unfortunate infestation of the lab by what seemed to be infinite population of mites.

Griffen was a true University of Texas student in the sense that he seemed utterly oblivious to the devastating effects of mites on Drosophila cultures and the place was overrun with them. Back in those days just about everybody smoked and it was unbelievable but true that when a cigarette was balanced on the edge of one of the lab tables in the fly lab there, the pinkish tinge to the table tops would quickly disappear around the cigarette, revealing the original black finish as the mites fled the cigarette heat and smoke. The first thing that a person would do on going home after working in that lab was to take a shower and change clothes—even then a person would have an itchy feeling.

This did not seem to bother Griffen who, as a Texas graduate, had learned to live gracefully with them. Among the workers in the lab at Missouri were Dan Lindsley, who was working for a master's degree at the time, and Katie Green, who, along with her husband Mel, had been a student of C. P. Oliver at the University of Minnesota. The latter realized that as fast as she was generating new chromosome aberrations from X-ray treatment, she was losing them to the mite onslaught. In desperation she came to me.

Katie asked for my help out of her dilemma of spinning the research wheels where it was obvious that nothing was being accomplished; I suggested that she might do some more work on the alleles at a locus called "lozenge" which showed the peculiarity of seeming to be located at slightly different positions on the chromosome, unlike most alleles which are usually assumed to be different mutational changes at the same locus. Mel (her husband) and C. P. Oliver had published a short note on this while he was a student at Minnesota and I knew that they had a continuing interest in this kind of work. (It should be noticed in passing that Ed Lewis, after transferring from Bucknell to the University of Minnesota also worked in the lab of C.P. Oliver, and noticed that the Star and Star-recessive mutants I had sent him also seemed to be located at different positions (Lewis 1939). This observation later formed the basis of his Ph. D. thesis.)

There was a problem, however. The genes they worked with at Minnesota were located within an inverted sequence which made it impossible to test them experimentally with others found on normal chromosomes. So Katie started doing things with those alleles that were workable and I took it upon myself to try to get the first two alleles, which served originally as the basis of the investigation, into normal chromosomes.

I tried the obvious well-established procedures—radiation and special genetic backgrounds—with no success. And then I got a brainstorm, one of the very best in my scientific career. Put in common terminology, the genetic problem basically centered on getting the simultaneous occurrence of two mutually exclusive events, either of which would kill the organism by itself but which, if they occurred together, would allow the organism to survive. As a simple-minded analogy imagine that to achieve a certain result it was necessary to expose the fly both to the searing heat of a furnace and to the cold of a near absolute zero temperature at the same time. Clearly these two things cannot be done simultaneously.

I finally figured out a way of persuading the organism into doing the apparently impossible. It amounted to finding a way of separating the two conflicting treatments, and finding the means for allowing the fly to survive the otherwise completely lethal treatments (Novitski 1950).

The genetic scheme for doing this, which involved recovering the complementary products of a crossover in a single egg, something that ordinarily never happens, this was remarkably successful. Although I was delighted, some were not. When I described this to L. J. Stadler, the renowned corn geneticist at the University of Missouri, he remarked "You Drosophila geneticists have the most complicated way of doing things" and changed the subject to more interesting topics. However, Curt Stern remarked that my brief paper in the journal Genetics on this was more like a chess game than any other paper he had ever read, and "represented the epitome of sophistication in a rather esoteric field." Sturtevant indicated in his laconic way that he thought it was an unusually ingenious solution to an apparently impossible problem. And so we finally get back to Muller.

After the war I spent time at Caltech as a postdoctoral fellow and during this time I heard that Muller (whom I had never met) was visiting Pasadena to see his son who was a student there. Although I was aware of the antipathy between Sturtevant and Muller, it was brought home when I found out that Muller would not be invited to the lab or to Sturtevant's house for an evening, which would have been customary for any visiting dignitary. It was also obvious that there were the elements of a slight involved. So I got in touch with Muller and invited him to have dinner with Esther and me at our house. I looked forward to discussing Drosophila problems with him, particularly one I had just finished working on.

So it was that after dinner I asked him how he would go about getting desirable alleles out of inverted chromosomes where crossing over was suppressed and ordinary experiments would not work, and how he responded by changing his pacing into a kind of child-like skipping. He clapped his hands over his head repeatedly in what I can describe only as pure scientific joy. I am as susceptible to praise and flattery (however well-deserved) as the next guy, so at that point he became my friend for life.

That Muller was impressed by my tour de force with getting genes out of inversions was brought home again some years later. I got a letter from Robert Graham who ran the original sperm bank in Los Angeles, founded by Muller and Graham, where selected males are asked to make contributions so that women concerned with the quality of the next generation can avail themselves of the best DNA possible for the paternal genes of their offspring. Graham thanked me for my cooperation. However, to allay the apprehensions of those who know me well and would be concerned to learn that I could make a disproportionate contribution to the next generation, I should hasten to say that despite Muller's apparent recommendation, I do not recall ever being asked to make an actual contribution, nor am I aware of any queues of women asking for same. I guess that they found enough volunteers amongst more obviously qualified candidates!

I can only conclude that my inclusion on this list was Muller's idea, which originated in our living room in Altadena. In any case, I was able to send to Mel and Katie Green a purified version of the two lozenge alleles, which they then could use in their analysis of the fine structure of the lozenge locus, if they wished.

The Sturtevant and Beadle Blunder

Early in my postgraduate career I became interested in the apparent stability of ring chromosomes in Drosophila. In my imagination I could see them replicating and getting entangled around each other so that they could not move easily to opposite cells during division. As a result they should occasionally be lost and this loss should be experimentally demonstrable. There existed a set of experiments with numerical results where this might be manifested, in the classic paper of Sturtevant and Beadle (1936) where they used a special chromosome to prove that single crossing over occurred freely in inversion heterozygotes, and, as discussed in Chapter Eight, incidentally provided the answer to one of the outstanding problems in population genetics at that time.

Examining their published data closely, however, proved to be unrewarding. Their observed numbers did not make any sense. My first reaction was to assume that there was some sort of printer's typographical error. I went to Sturtevant and asked him about the disagreement of their data with their analysis. For one thing, the experiment allowed for two independent measures of the production of ring chromosomes and the two were not in agreement. Second, the figures they gave showed that the amount of crossing over was well over 100% although the method of analysis precluded any value over that amount. Third, the figure they gave for the calculated amount of crossing over did not agree with their observed numbers. Finally, the chromosome they used in their experiment, one which they had to synthesize, was primitively constructed with predictable effects decreasing viability, yet they were getting too many, not too few, ring chromosomes.

Curiously, now some ten years after receiving my Ph. D. with Sturtevant as my thesis advisor, this was the first time that he and I together were to confront a genetics problem involving the fundamentals of Drosophila genetics, and that was in a discussion that lasted no more than a minute! Throughout the entire graduate career the emphasis had been on population problems (20-1). Sturtevant had never discussed his work or the work of others in this area, either in class lectures, in informal discussions or in seminars; his expositions had been concerned with history. Now my research career was about to take an unexpected turn because I discovered what I first thought was a typographical mishmash.

Sturtevant's response to my query about the incompatibility of their numerical results and their interpretation and was one of surprise. "Well," he said, "those experiments were done by Beadle and you will have to ask him." So I did. Beadle's response was that he did not remember. (This was understandable. It had been more than ten years since the experiments had been done.) He did recall, however, that someone in England had written him asking to see the data, which were then sent off to England. Apparently I was not alone in wondering about this experiment. He did not remember who it was.

The first problem was that he derived two different measures of crossing over from his experiment. One was 116% and the other was 72%. But he gave, as a final result 86%. How did he arrive at this figure? After some cogitation it occurred to me that, of course, he was bothered by the 116%, how could it be greater than 100%? So apparently he changed the 116% to 100% and averaged it with 72% to get 86%! In this way he avoided the conclusion, which would have been troubling, that he was dealing with a hitherto unknown phenomenon. In any case, I now had the conviction that the figures in as given in the paper were not typographical errors.

So I was left with the feeling that the matter should be reinvestigated, and in a more acceptable way. When Beadle (almost certainly under Sturtevant's tutelage) undertook the problem of making a chromosome that would spontaneously manufacture ring chromosomes, he apparently took his basic material off the stock shelves in the Caltech stock collection or, possibly, directly from Calvin Bridges. The result was a "quick and dirty" solution that produced a highly abnormal ring chromosome with extensively duplicated material that killed all male progeny that carried it; females survived but their relative viability must have been affected as well.

I decided to repeat their experiment, but to do it in a more leisurely fashion, starting with a ring of good viability in both sexes, by opening it out and then letting it produce like rings similar to the original by crossing over. With Dan Lindsley (Novitski and Lindsley 1950) we finally succeeded, and were able to confirm that the Beadle numerical results, although inexplicable, were valid and repeatable.

Further experiments showed that there was indeed a new phenomenon involved, one that operated in the formation of the egg in the Drosophila female, that when there were two structurally dissimilar chromosome strands competing for inclusion in the egg (the other going into the polar body), the smaller tended to be included. This I called "nonrandom disjunction" (Novitski 1951).

Interestingly, once that phenomenon had been established, it was not difficult to find cases in the already published literature where that effect had confounded previous workers. Indeed, in that same paper of Sturtevant and Beadle, where they measure the amount of crossing over between two chromosomes which happen to be structurally dissimilar, they record a 2:1 ratio among the crossovers instead of the usual 1:1, but they fail to mention anything remarkable about that result. This is understandable. Their work was so extensive, including many different combinations of chromosome types, that the anomaly was buried in the mass of data presented. Mather (1939) remarked that in one of his experiments a "deficiency is recovered twice as frequently as the corresponding portion of the other chromosomes . . . . [but] it seems to be bound up with a maternal effect." The most extensive study in which nonrandom disjunction was missed was that of Glass (1935) who, in a Ph. D. thesis supervised by H. J. Muller, concluded that lethal genes were responsible for the discrepancy between the complementary classes after crossing over occurred in translocation heterozygotes. S. Zimmering (1955) in a thoughtful repetition of that work, showed conclusively that Glass's (and therefore Muller's) interpretation of a deviation based on the effects of deleterious or lethal genes was incorrect, and that nonrandom disjunction was responsible.

My reaction to the repeated oversights (by Sturtevant, Beadle, Muller) of this simple phenomenon by attributing abnormal ratios to lethality or some other trivial disturbance led to a determination on my part never, under any circumstances, to accept inviability as an explanation for unusual experimental results until all other possibilities had been exhausted. This determination later led to possibly the most serious of all the mistakes during my experimental career. This will be discussed later.

Bernstein, Sturtevant and the ABO Blood Groups

Every worker in science with any intelligence or imagination (or both) will, from time to time, commit some error, some misjudgment, that in retrospect seems somewhat stupid or foolish. The writer himself can lay claim to a few such. These will vary in importance from inconsequential to egregious, but few will reach the level of incomprehensibility as one made by the eminent mathematician Felix Bernstein in his classic analysis of the ABO blood groups.

Beginning students who have been exposed to elementary genetics and are accustomed to thinking of genetic analysis in humans as based entirely on the study of pedigrees are always surprised to learn that valid genetic information can be obtained from an examination of the frequencies of types appearing in a population, without regard to family relationships.

In teaching an elementary course I was faced with the problem of presenting this concept in a simple way, but the only formal presentation, one which I had used in advanced classes for years, was based on the classic paper of Bernstein (1925) who, after a rather extensive series of mathematical manipulations, showed that the population frequencies of the four types, A, B, AB, and O, were incompatible with the theory that they were based on two genes, A and B, located on different chromosomes. Instead he proposed that they were simply two different forms of the same gene, two alleles, where a given chromosome could carry one or the other (A or B) or neither (O) but not both. This analysis was given wide acceptance by its presentation in the first really comprehensive human genetics text, that of Curt Stern (1949).

I wanted to use data from the frequencies of the ABO blood groups, but to do so without the complications of the algebraic manipulations provided by Bernstein. How to do this? Then I recalled a statement made by Sturtevant during a graduate course at Caltech many years ago (1939), as he finished his presentation of the analysis of Bernstein. "We knew that the multiple locus theory of the ABO groups was wrong but we didn't know what the correct theory was until Bernstein published his paper." It apparently never occurred to those of us in that class to ask how, in those early days of genetics, he knew that the older theory was incorrect although, from the casual way in which the statement was made, he seemed to imply that the reason was obvious and that any further explanation was unnecessary. Unfortunately Sturtevant had passed away by the time that this became an issue to me.

So how did Sturtevant know that the multiple locus theory was wrong? Knowing his preference for simple solutions, I knew that the answer must embody the easily understandable explanation I was seeking for use in an elementary class. After some concentration the lamp in the cranium, which had been flickering dimly, suddenly went bright. Of course. The answer was obvious. He had simply applied the multiplicative law of probability: that the probability of two independents occurring simultaneously is equal to the product of their individual probabilities. Thus, if the chance of heads after tossing a coin is 1/2, then the chance that two heads will come up in two tosses is 1/2 x 1/2 or 1/4.

Sturtevant must have argued that if the A and B factors were independent, then the frequency of those carrying both factors should equal the product of the frequency of all those carrying the A factor times the frequency of all those carrying B. But the data do not bear this out, in fact the observed frequency of those carrying both is only about half as great as that expected. So this is undoubtedly how Sturtevant "knew that there was something wrong."

However, the surprise came upon reexamination of Bernstein's conclusion. After a couple of pages of intricate algebra, his concluding formula used to show the incompatibility of the theory of two independent factors turned out to be nothing more than a statement of the simple law of probability (Novitski 1976). The assumptions that went into the formulation of his mathematical analysis all were obviously unnecessary, and he could have replaced his entire paper with one sentence. How could such a competent mathematician as Bernstein (for his considerable mathematical contributions, see review in Crow 1993) make such a blunder?

My satisfaction at having made some small contribution to this field was blunted, however, when I received a communication from Alexander Wiener, one of the world's leading authorities in this field, saying that he too had published the same observation about Bernstein's analysis some years earlier, and implied (somewhat testily, I thought) that I should become more familiar with the literature in journals of hematology. I still haven't found his article. Interestingly, I found my position reversed in my discussions with Leo Szilard, where I was the one whose publications were being overlooked (Szilard 1960).

After his elaborate (and unnecessary) algebraic exercise, Bernstein shows that another formula, one based on the assumption that the two factors, A and B, are simply different forms of the same gene (alleles) fits the observed data quite well. This had been accepted by me, by Sturtevant, Stern and undoubtedly dozens of others who taught courses beyond the elementary level, as adequate. However, after a lecture that included Bernstein's proposal for allelism, a student came to me and complained. "I am not convinced," she said. "How do we know that there is not another formula that fits equally well, or maybe even better?"

And she was right, of course. My presentation lacked any scientific validity. I was presenting as scientific fact something that I expected the student to take on faith, more appropriate in a theological presentation than in a scientific discussion. How could I satisfy the student (and myself) in a simple way that the supposition that the two factors were allelic and not located on different (i.e., independently segregating) chromosomes was valid? After some cogitation, and a little arithmetical experimentation I decided that I could.

The procedure was to assume that the three genetic factors A, B, and O occurred randomly in the germ cells and to calculate the frequencies of the four possible combinations A, B, O and AB, using the known frequencies of their combinations in individuals in the population. It immediately becomes clear that the frequency of germ cells carrying AB must be zero. (see Novitski 1982, pp. 403-407). Then the student is forced to the conclusion that A and B, when present in an AB individual, are always found separately on two like chromosomes (homologs) and always separate from each other at germ cell formation (meiosis). Thus instead of *assuming* that the factors are alleles, as Bernstein seems to have done, this procedure forces one to *conclude* that they are alleles. This is such a simple and convincing argument, just the sort that Sturtevant would do, that one wonders why he didn't. The best answer is that he probably did, but at a time early in the development of human genetics when the data were fragmentary and contradictory, and proved to be inconclusive, and that he never bothered to recheck after better data were available.

As for Bernstein, it is my guess that after discovering that the purely algebraic demonstration led to equations of high degree that were inherently insoluble, he undoubtedly made precisely this sort of calculation but being a professional mathematician with some personal pride, was unwilling to admit the algebra involved was too cumbersome to solve or to admit that his conclusion of three alleles at one locus was based on simple arithmetic without recourse to any mathematics others than the addition, subtraction and multiplication of crude numbers.

The X-Y Chromosome

When Dan Lindsley came to Caltech for his Ph. D. work, I had suggested that, for his thesis problem, he study the composition of the Y-chromosome by examining closely the products of exchange between the X and Y chromosome. Now having finished his thesis, he was in possession of a number of unusual chromosomes, combinations of the X and known pieces of the Y chromosome.

Of course, combinations of pieces of the X with pieces of the Y were very well known, having been the main thrust of the Ph. D. thesis work of Curt Stern in the late nineteen twenties in Germany, followed by a continuation of this work for several years after his arrival in this country. After the discovery of the action of X-rays on the breakage of chromosomes, Muller succeeded in producing many new types of chromosome combinations, some involving the X and Y chromosomes.

However, all these earlier efforts produced pieces produced by random fragmentation rather than synthesis of specific desirable types, and our interest lay in the production of a single chromosome that would carry all the necessary genes of both the X and the Y chromosomes.

Of course we had no way of knowing whether such a single chromosome (technically a "univalent") would allow for fertility of the male, or whether it would successfully go through the meiotic divisions which produced the mature sperm. Although there might have been several ways to approach this difficult problem, the final scheme worked out took advantage of two unusual chromosomes, one from Lindsley and one from me.

For my part, I had stumbled upon an unusual chromosome in my effort to repeat the Sturtevant and Beadle incomprehensible ring chromosome experiment. To do a better job, I needed a ring chromosome of good viability in both sexes. There was one labeled as such in the stock center at Caltech. However, on testing, it proved not to be a ring but to be a rod shaped chromosome, in inverted sequence (21-2). Logic suggested that such an opened out ring would have a tuft of nonessential material at its tip and this was verified by cytological examination.

The method adopted to produce the chromosome had several desirable features. First, it involved no irradiation (which might mess up the integrity of the chromosomes involved) and, second, it depended on an event that would occur in the male spontaneously and thus might more clearly represent the product of the pairing of those two chromosomes. This would maximize the likelihood that any product recovered would retain intact the pairing configuration of the originally independent chromosomes.

The final experimental design was a dream come true, the most efficient, elegant experiment ever designed (in my humble unbiased opinion, of course) in all of Drosophila genetics. I can say this without undue personal embarrassment because my recollection (perhaps faulty after all these years) is that much if not most of the design came from Dan Lindsley.

The procedure was to put into males two independent chromosome pieces, both of which were essential for the fertility of the male, but which would regularly pair and separate from each other at germ cell formation. Male progeny would then receive an incomplete set and would be sterile. Exceptions would consist of those male progeny which received the genes of both chromosomes, which could happen in one of two ways. In rare cases the two chromosomes might fail to separate (nondisjunction) and could appear together in a sperm, giving rise to a fertile male. Or they might undergo a spontaneous attachment, so that, now bound together, the X-chromosome had all the fertility factors of the Y-chromosome attached to it, i. e., our desired X-Y chromosome.

In this system, since ordinarily all males are sterile, the progeny of an initial cross could be dumped into a fresh culture bottle which was later examined for signs of progeny. So hundreds of males could be tested in each culture without looking at a single individual, because virtually all cultures would be completely sterile. It involved little effort to set up hundreds of such cultures which meant that many thousands of flies could be tested without looking at a single one (21-3).

Further, the crosses were arranged so that when exceptional fertile males made an appearance, the trivial unwanted cases of nondisjunction gave only sterile males among their progeny whereas the very rare case of bona fide attachment, if they occurred, would produce fertile males, and perpetuate the stock.

Of the many tens of thousands of males that tested themselves according to this plan, there were four fertile ones. Three proved to be nondisjunctional (at least their male progeny were sterile) but the fourth produced fertile male progeny and so was the sought after chromosome, one with all the essential genes of the X-chromosome as well as those of the Y-chromosome (Lindsley and Novitski 1950).

Cooper used this chromosome in his cytological studies (Cooper 1959) and we had, not only a chromosome extremely valuable for future work, but also the great pleasure of having performed an experiment which in its planning, execution and success greatly exceeded the norm for the usual run-of-the-mill work with Drosophila.

My Most Notable Error

For every success in science there are several failures (at least that has been true in my career). One particularly notable case occurred in the study of chromosome behavior in the Drosophila melanogaster male.

It was caused in part by the failure of some of the more competent workers in the field (Sturtevant, Beadle, Muller) to recognize nonrandom disjunction in their experiments when it was staring them in the face, and by their subsequent willingness to attribute their anomalous results to a catchall triviality, differential viability, which means simply that individuals of one genetic constitution are more viable than those of another. Thus, the experimental observation of a difference in two frequencies when equality is expected can be dismissed as an inevitable result when one class of fly is "healthier" than another.

In some simple experiments performed for a master's thesis, Iris Sandler got a set of results where four numbers, instead of occurring as two sets of equal pairs, appeared to be random. After several hours of cogitation, it occurred to us that in fact there was a simple relationship. The genetics are bit difficult, but as an analogy for this, consider that a rich uncle has left one of his heirs a bit of property and has stated specifically the GPS positions of each of the four corners. The numbers make no particular sense until the beneficiary calculates the distance between the corners and realizes that adjacent corners are equidistant, and that the two diagonals also equal each other. The conclusion is obvious: the plot of land is a perfect square.

Thus it was that the experimental numbers from this Drosophila experiment, when calculated "diagonally" were internally consistent. From this I concluded that such a mathematically precise result could not come from "differential viability" but must reflect a phenomenon occurring regularly during chromosome segregation in the male. At this point, wherever I looked at the data published Drosophila experiments, or whenever I made additional observations, I could see manifestations of this new phenomenon in Drosophila.

However, I was wrong, although it took the work of many others over a period of time to convince me. Although many in our laboratory were sure that my idea was faulty, there was no really irrefutable evidence until a paper by Hartl, Hiraizumi and Crow (1967) showed beyond any reasonable doubt that I must be on the wrong track. It turned out that differential viability occurring at the level of the sperm cell was, after all, the correct explanation.

In cases like this, it is quite unnecessary to publish a retraction. Others, particularly close friends, are not reluctant to point out, in print or otherwise, and I might say, somewhat gleefully, that a serious mistake had been made!

The Deadly Final Oral Ph. D. examination

Programs for advanced degrees usually terminate with an oral examination of the candidate, who undergoes a grilling on the content of the thesis and particularly on its conclusions. These are important to the student and his advisor who consider that the final outcome may determine the student's future more than any other single event in the academic life of the student. In practice, other events, such as extensive written exams, general appraisal of the adequacy of the research, etc., may have preordained success (or failure) in the effort to get a higher degree. In fact, most advisors will not allow the student to advance to the final exam if there is any doubt as to its eventual positive outcome.

But such exams demand a committee of evaluation and for the members of the committee the required exam can be a real pain. They are usually presented a copy of the thesis beforehand for perusal. In fact, theses can be extremely voluminous and correspondingly boring. Although neophyte faculty will try to fulfill their obligations, most older faculty will barely glance at the thesis but will await the final exam to hear the candidates describe their work and present their conclusions, after which they may feel free to ask any questions that may seem appropriate (and some that may not). Every faculty member can relate instances where this informal ad hoc procedure has had unintended, sometimes humorous, consequences.

At an exam for Larry Sandler at the University of Missouri, the committee included Ernie Sears, a well-known cytologist who had distinguished himself for his work on wheat. At the conclusion, when Sears was given the opportunity to ask questions, he weighed in with some pretty heavy stuff, material which even an expert in the field of plant cytology would have problems with. Although Larry was able to handle a few, mostly he was at a loss and had to parry with good humor and friendliness. For a while I feared for the outcome but finally Sears acknowledged that the performance was acceptable.

Walking out afterwards, I remarked to Sears that I thought that his expectations seemed to me to be a bit excessive for a beginning student just getting a master's degree. "Oh," said Sears, "I thought that was a Ph. D. exam."

Apparently he had not only not read the thesis, which as an accomplished plant cytologist he would not have understood anyway, he hadn't even read the title page which clearly stated that the thesis was presented for a master's degree.

One case in which a committee at Berkeley was clearly asleep was one that I did not attend, but which I became aware of when the editors of a journal sent me a paper for review before publication. It involved the phenomenon known as somatic crossing over.

Crossing over in somatic (body) cells has consequences quite different from those produced when ordinary crossing over occurs in germ cell formation. In the latter case, if one chromosome carries two genes at different loci, *A* and *B*, and it pairing partner, its pairing partner (homolog) carries *a* and *b*, then a crossover will produce two new types, one chromosome with *A* and *b* and its complement with *a* and *B*. Crossing over in somatic cells, however, has other consequences and even the experienced geneticist has to resort to diagrams to understand them. Stern (1936) did the most detailed pioneering work on this phenomenon, initially with the thought that it must play an important role in development. As it turned out, it is a less than fundamental phenomenon although it has been implicated in humans in disease (retinoblastoma, for instance).

Among the peculiarities of somatic crossing over is the fact that its incidence can be greatly increased by the presence of a special class of mutant genes called "Minutes." These genes generally decrease the size of the flies bristles, at the same time prolonging the life cycle somewhat. Curiously, their effect in increasing the rate of somatic crossing over had been found only when the location of the specific Minute being investigated is on the same chromosome arm as the crossover.

For his Ph. D. problem, one of Stern's graduate students undertook a study to see if he could make more sense of this intriguing problem. After several years' work, he presented his thesis to his final committee at Berkeley. They agreed that it represented a creditable piece of research and recommended that the degree be granted, which was an appropriate decision. However, as I read the paper being presented for publication, I realized that there was a problem. Instead of throwing light on the phenomenon his results actually made it more confusing. The earlier rules were not being followed in his experiments and the results, all apparently bona fide, simply made no sense.

As I pondered this quandary, the lamp in the skull suddenly grew brighter. Could it be that he had simply accidentally switched the labels on two of his standard stocks used for the tests? With pencil and paper I quickly verified that this was indeed a reasonable explanation, and then I dashed off a note to Stern suggesting that the two stocks be tested. The tests were eventually conducted by Stan Zimmering who showed that in fact the two stocks were mislabeled.

The student (whose name I have forgotten) made the necessary alterations in his manuscript and resubmitted it for publication. Fortunately his earlier work was of sufficiently high caliber that the decision to approve the thesis was still valid. But where were the minds and critical faculties of the distinguished members of that examining committee at Berkeley before and during that exam? Undoubtedly they were intellectually sound asleep, as I myself have been on a number of such occasions.

Curt Stern, however, seemed to have been impressed. When, some time later, he decided that his work on gene action in development would be advanced by the use of special chromosomes of a type never seen before in any experimental organism, he asked if I could provide it. At its very tip, beyond any of the normal genes found on that chromosome, this hypothetical new chromosome was to have an extraneous genetic marker readily visible in the development of the fly. This was a challenge and Stern knew that I would rise to the bait. It took more than a year before I was able to provide him with several different cases; subsequent tests with DNA probes showed that in fact we had succeeded in cutting into the terminus (the telomeric region) of the chromosomes involved. These unique chromosomes also provided an opportunity to make some unusual combinations that were useful in the study of meiosis. In retrospect, it is interesting to note that this kind of chromosomal attachment bears some of the same characteristics as the one major difference in chromosome structure between human and the chimpanzees, in which two chromosomes normally separate in the chimpanzee are attached at their telomere regions in humans (See Appendix D).

New Thoughts about Mendel

The classic work of Gregor Mendel on inheritance in the garden pea, Pisum sativum, has been the object of greater scrutiny by more people in more diverse professions that any other bona fide scientific work. These can be broadly separated into two areas: the examination of Mendel's numerical results and the broader implications of his thoughts on biological phenomena. These works appear in journals in printed form but are also now available as stand-alone presentations on websites, and are so extensive that a close examination of all would require the equivalent of several Ph. D. theses.

The work in the area of numerical results came to pass in large part because of the analysis by the English statistician R. A. Fisher, who concluded that Mendel's experiments, viewed as a whole, came out numerically too close to ideal expectations, violating the usual rule of some statistical variability. Broadly speaking, two camps emerged: on the one hand those who felt that Mendel was being maligned by the implication that he had cheated, and on the other those who felt Fisher was being justifiably critical. In some cases the defense of Mendel rested on the assumption that a man of the cloth, as he was, would not engage in petty deception. One person has even suggested that perhaps his great success had its origins in divine guidance. This controversy has gone on for years and will, no doubt, continue for many more.

As a student of genetics, I was of course aware of Mendel's contribution but paid more attention to it when I began teaching a course in elementary genetics. At that time I searched for a suitable text and perused several dozen texts in the field available to me, sent by publishers.

I became aware that more than half of all elementary texts presented his work so badly that they were in serious error, indicating that the authors themselves did not understand what Mendel had done. Perhaps this shouldn't have been surprising. Back in those days (1970s), biology textbooks were usually written by teachers with a naturalistic bent who may have had a greater or lesser deficiency in what we have been describing as a logical-mathematical intelligence.

The major misrepresentation in those texts was that Mendel's predominant results, those involving the seed characters round vs. wrinkled and yellow vs. green, were presented as inherited characters, like, say, albinism in humans. Apparently the authors were oblivious of the fact that these seed characters, present individually in the thirty or so seeds on each plant, in fact represented the next generation, making it unnecessary to plant each seed and wait for the appearance of the mature plant to make a determination of its characteristics. This simple fact made it possible for Mendel to get numerical results on a grander scale than would have been possible with simply inherited plant characters. To put it more dramatically, his experiments involving seed characters were at least 3000% more efficient than those with plant characters, and even more so if one considers that he was relieved of the necessity of planting seeds and growing plants to see the distribution of characters in the next generation. Thus Mendel could get sizeable numbers in a very constricted garden space, and in a shortened period of time. This elegant aspect of Mendel's experiments was being completely omitted in many of these elementary textbooks and the impact that this might have had on the more adept student was completely lost.

There were also the occasional gross misstatements, fortunately rare, that accidentally plague even the most scholarly of works. A beautiful example of this is found in one of the best biology texts of the time that Mendel did his work on the sweet pea, Lathyrus odoratus. This points out another feature of the many works about Mendel. Of the many, perhaps hundreds, who have ventured an opinion about Mendel's paper, few—probably not more than a few dozen—ever took a close look at the garden pea in the field and even fewer ever grew the plant to see its development. I can say this with some authority since I myself belong in that class. It was only after I decided that Fisher was undoubtedly mistaken in his characterization of some of Mendel's experiments that I realized that I needed to know more about the seed characters. I sought the help of a bona fide pea geneticist, Prof. James Myers of Oregon State University, and only then came to know that the commonly described seed characters of round vs. wrinkled and yellow vs. green did not show up conspicuously as they were pictured in all texts, but required the skillful removal of the outer opaque seed coat with a thumbnail to reveal the seed characters hidden underneath.

One of the more irksome misstatements sometimes found in those texts went something like this: "Mendel was indeed lucky to have limited his studies to seven characters, because the garden pea has only seven chromosomes, and if he had studied eight characters, two would have been on the same chromosome, would have not shown independent assortment, would have confused his results, and probably would have prevented him from writing his paper."

Anyone with a knowledge of elementary genetics would realize that such a statement originates in a confusion of the concepts of "linkage" and "synteny." Linkage refers to genetic results indicating an association of two or more characters in an experimental setup, while synteny means the actual physical location of the loci of two or more characters on the same chromosome. The two are far from synonymous. Apparent linkage can be found when chromosome aberrations are present, and synteny is not revealed by experimental tests when the loci are sufficiently distant that the ubiquitous presence of crossing over separates the different genes. Furthermore, any statement of the sort quoted above would require specific tests, usually extensive backcrosses, of all 21 combinations of seven characters taken two at a time.

I myself was incompetent to carry the argument further and I inquired of my good friend Alexander Fabergé the name of a pea geneticist who may have published in this area. His response was that the best work was done in Sweden, and that one of the authorities there was Styg Blixt. I wrote to Blixt pointing out that while information might be available in Sweden (published in Swedish or German) about the chromosomal distribution of Mendel's characters, it was not generally available in the United States. Would he be willing to write a short note for publication in the English language in an effort to educate non-pea geneticists in this country? He wrote back indicating his willingness to do this.

Many months passed without any more communication from Blixt. In the meantime I arranged to spend a sabbatical in the Netherlands. University faculty on sabbatical ordinarily are on half-salary and usually try to find the missing half-salary from other sources. Because of the stiff competition for outside funds, particularly from those competent but prestige conscious faculty from schools all over the United States who would like to associate with colleagues in Oxford or Cambridge, or whose families would insist on certain amenities, like the art and history of Paris or Rome, or the language of London, we decided that the chance of approval would be heightened if we chose a suitable but less fashionable destination. Previous contacts with Prof. Sobels at the University of Leiden suggested the Netherlands. To further increase the chance of adequate funding (in the vernacular, to hedge my bets) I applied to the Fulbright Foundation, the John Simon Guggenheim Foundation, and the National Science Foundation, with the hope that at least one would be successful. As it turned out, my hedging worked out better than expected, all three requests were approved, so that with four half-time salaries, I was able to extend my sabbatical to two years.

Thus it was that my interest in Mendel's work received a boost by proximity to the main source of information. Adding to this was a curious feature of our apartment in Holland. Sobels was not able to find us living quarters close to the University, but did find a nice spacious apartment in a high-rise building overlooking the North Sea, in a village called Nordwijk-an-See. The apartment was owned by a German woman who made a small fortune each summer by renting it out at outrageous rates to German tourists who wanted to spend a few vacation days on the Dutch coast. So our lease carried the provision that we had to leave for the summer months and we arranged, happily, to spend those months at the University of Turku in Finland, passing through Lund, Sweden, on the way there and back, where we stayed each summer for a week or two.

Thus it was that I was able to visit the plant breeding institute at Landskrona and make the acquaintance of Styg Blixt, hoping to spur him on to write the paper he had promised. It turned out, however, that he had not only written an account, but had already sent the printer's proofs back to the journal Nature. My initial reaction of pleasure turned to one of disappointment when I read his account and realized, even with my limited knowledge, that he had made a substantial error in stating that Mendel had not made the crosses that might reveal linkage, when, in fact, he had. (Interestingly, one combination where Mendel might have run into linkage was made extensively, not for scientific purposes, but because the seeds were exceptionally tasty!)

My discomfort with Blixt's account was heightened after conversations with another pea geneticist, Nilsson, at Lund University, who, along with two other well-known Mendel authorities, A. Gustafsson and R. Lamm, insisted that Blixt's assignment of the loci of three characters to one chromosome was incorrect, and that instead two chromosomes each carried two characters far enough apart so that no evidence for linkage appeared in the relatively low numbers in those cases, with the three remaining each positioned on a different chromosome.

What had started out as a casual attempt to get into the printed record, in English, a statement that all seven of Mendel's characters probably did not occur each on a different chromosome, had inflamed a standing controversy among the pea geneticists of Sweden. Thus my initial suggestion to Blixt that he write this simple paper had inadvertently put him in a very awkward position. I offered to try to resolve the differences by writing a more detailed account outlining the two points of view. Blixt agreed, and the result was the joint paper reproduced at the end of this account.

Several years later, in an independent effort with Prof. Raphael Falk of the Hebrew University in Jerusalem, who was visiting Oregon as a research professor, we got the idea that it might be interesting to see if a computer could duplicate a carefully chosen set of Mendel's experiments. A computer is an ideal mechanism for such a check because it can simulate the biological process of meiosis perfectly. This is done by generating random numbers between 0 and 1, and selecting those with a value between 0 and 0.5 to represent one allele, say A, and numbers between 0.5 and 1 to represent the other, say a. It can then reproduce the various classes of progeny up to Mendel's total in a given experiment many thousands of times a minute, and Mendel's results can then be compared with the computer-derived distribution.

Another great advantage of a computer simulation is that the production of a curve of distributions of many thousands of results makes it possible to see where Mendel's results fit in, without recourse to statistical techniques which themselves have been held up to question when applied to Mendel's results.

Preliminary runs showed that the computer was having some difficulty getting results as good as Mendel's. Here the problem languished until it was taken up by Prof. C. E. Novitski (better known to me as my son Charles) who greatly expanded the scope of the comparisons and concluded, among other things, that Mendel probably could not have achieved his results by running duplicate sets and subsequently selecting those sets which gave the best results. His data simply did not have the number of cases of considerable deviation (deviations which seem unlikely to the average observer) but actually are statistically expected.

My next exposure to Mendel came when Charles reported to me that Sturtevant, in his History of Genetics, had inadvertently misrepresented the probability of some of Mendel's results. These refer to certain experiments which have been described by Wright (1966) as "the most serious evidence for fraud by Mendel as presented by Fisher."

The two sets of experiments in question are those in which progeny tests may be required to determine whether certain plants carried a specific recessive allele. Fisher had pointed out that when such tests are made to distinguish a homozygote (AA) from a heterozygote (Aa), testing for the possible presence of the a allele by selfing the plant and then planting a limited number of seeds, say ten, a fraction of the tests will fail to reveal the presence of the recessive allele. Simply by chance the result of an Aa x Aa mating (actually a selfing) will fail to give even one homozygous aa progeny. Fisher calculated that between five and six percent of such tests should fail for this reason, but Mendel's figures not only showed no such discrepancy, but were closer to an exact two to one ratio that he might have expected. How did this come about?

The original statement, as it appeared in Mendel's paper, described the experiment in the following way: "Für jeden einzelnen von den nachfolgenden Versuchen wurden 100 pflanzen ausgewählt, welche in der ersten Generation das dominirende Merkmal besassen, und um die Bedeutung desselben zu prüfen, von jeder 10 Samen angebaut." The exact meaning of this sentence has been the subject of some controversy.

The Batesonian translation—the commonly accepted one—is: "For each separate trial in the following experiments, 100 plants were selected which displayed the dominant character in the first generation, and in order to ascertain the significance of this, ten seeds of each were cultivated."

In this initial version of Bateson, the last word "angebaut" was translated as "cultivated." It does appear to be awkward in the sense that the word "cultivate" ordinarily suggests some positive action that improves a condition, and how this might apply to the garden pea seed itself is not clear. It leaves open the possibility that more than ten seeds were planted, and, from the seedlings that appeared, ten were selected and nurtured (i.e., cultivated). In fact, Fisher, and Sturtevant after him, suggest that Mendel could have raised more than ten plants, thus limiting the percentage of errors that he would have made in his determinations. It should be noted, however, that Fisher states, when dismissing this possibility, as well as the one where Mendel might have made a different kind of test (a backcross), that Mendel's "style throughout suggests that he be taken entirely literally."

However, another translation, one by Sherwood, that appeared in a book by Stern and Sherwood (1966), translates "angebaut" as "sown." This translation makes more sense even though it is not the usual translation. Stern, although German by birth, was completely fluent in English and, because this translation appeared in his book, undoubtedly supervised any changes Sherwood made to Bateson's original translation. He may have recognized the slight awkwardness in the concept of cultivating seeds and must have preferred the word "sown." Further, the use of this term limits the actual procedure, since the suggestion that more than ten seeds were planted is obviated. Finally, Stern must have realized that this translation was directly opposed to the suggestion of his lifetime friend and scientific colleague Sturtevant that perhaps more than ten seeds were planted and more than ten plants raised (25-1).

Sturtevant, apparently taking his figure of 1/2000 from one of Fisher's tables, states that the probability of Mendel getting this result of only once in two thousand trials for these two experiments was a result so unlikely that it could be interpreted as a clear indication that Mendel was altering his data to fit his preconceptions. I myself had accepted this figure as valid and had used it in lectures as evidence in favor of Fisher's alteration hypothesis. Charles, however, pointed out that the figure of one in two thousand used by Fisher referred to all the experiments done by Mendel in the year 1863, and that the correct probability for the relevant experiment in 1863 was only about one in twenty!

Eddie Lewis had just promoted the reprinting of Sturtevant's book "A History of Genetics" and, feeling that this error found by Charles should somehow become part of the scientific record, I wrote Lewis about it. After some time I got a reply in which he agreed that Sturtevant had erred, but then went on to say:

"Mendel had the wrong expectation as Fisher showed—it is an ascertainment problem that Mendel did not recognize of course. Why others don't know this is because they have not studied Fisher carefully as Sturt did."

Although this statement was made in defense of Sturtevant, I regarded it as both a criticism and a challenge. I myself had not really looked critically at either the Mendel or the Fisher papers from the standpoint of their probably being incorrect, misinterpreted, or mistranslated in some important way, so I decided to read the two more carefully. I think that I was quite successful in discovering aspects of Mendel's and Fisher's works that had not been obvious to me previously.

Looked at from the standpoint of an experimentalist, it seemed clear that the controversial phrase must be looked at more as a summary of results than as a description of procedures. It obviously is not possible to raise precisely 100 plants in each of six experiments, to cultivate (or sow) exactly ten seeds from each of the 600 plants with complete success in every case. This would require a horticultural achievement beyond human capabilities. There must have been some failures, and although Mendel does not indicate the rate of failure (of germination, of producing seeds, or other causes) for these tests, he does give it for a subsequent experiment at about two percent. Ordinarily one pays little attention to failures in most experiments because they do not affect the overall outcome indicated by the successes. However, in Mendel's experiments they could have a profound effect.

Obviously, at the outset, Mendel must have thought that he would be able to classify with some certainty each dominant-bearing plant as either AA or Aa on the basis of the appearance (or not) of aa progeny after selfing, and must have considered ten to be the sufficient number for this determination. Any test with a number smaller than ten would be considered inadequate and would be disregarded, unless of course homozygous recessive progeny appeared, unambiguously identifying the parent as a heterozygote. Those so eliminated would be primarily of the AA constitution. Following this thread of thought through to its logical conclusion, the failure to detect some heterozygotes would be offset by the loss of some homozygotes because of the requirement of ten plants required for the determination (unless of course a homozygous recessive appeared, removing any doubt as to the heterozygosity of the parent). A crude calculation shows that these two opposing forces almost exactly balance each other (E. Novitski 2004) and a more elegant mathematical analysis by C. E. Novitski (2004), based on Mendel's failure rate of two percent, suggests that the expected ratio in this experiment is very close to the ideal of 2:1 (actually 2.07:1) and not the 1.7:1 that Fisher thought appropriate.

The next peculiarity of that statement is that he "selected 100 plants" for testing. It seems likely that he chose the number 100 because he could immediately see whether the results from any test deviated from the expected 67:33 (or 2:1 ratio) which his seed experiments had already shown to be the expectation. In fact, he promptly repeated an experiment he considered his most deviating (60:40) and got a ratio of 65:35, satisfying himself that all five plant characters were behaving precisely as the more extensively tested seed characters had previously predicted. How one interprets the word "selected" is not clear; one would hope that the selection of 100 from a somewhat larger number was not carried out with the ideal expectation in mind. The final total of 399:201 for all six experiments is so close to the perfect of 400:200 that some might feel uneasy about it.

If Fisher may have overlooked some aspects in the preceding experiment, he might also have erred in his evaluation of a subsequent experiment in which three characters were being followed, one of which required progeny testing and also gave results closer to the ideal of 2:1 than a 1.7:1 that Fisher predicted Two of the three characters are the old standbys of round-wrinkled and yellow-green. The third character (according to Fisher), the one in question, is that of colored vs. white flower color which, once again, would require progeny testing. But an examination of his numbers leaves one wondering.

Mendel gives the following figures for the trifactorial experiment: of 687 seeds planted in the first generation, 639 of the resulting plants fruited. Of these, 166 plants had white flowers and were unambiguously homozygous recessive, so their genetic constitution could be immediately determined. There remained 473, which could be either homozygous or heterozygous. Fisher suggests that Mendel raised ten plants from each of these, totaling 4730. Once again we may marvel at his skill and luck, in having 473 that needed determination and not having a single failure. Is it possible that each of the 639 "fruiting" plants were themselves the final stage of the experiment, that all the necessary information was provided by the seeds themselves?

The suggestion here is that Mendel did not make 4730 tests, or need to, because he used a characteristic of the flower color gene which made the heterozygote immediately visible: the coloration of the seed coat. He himself states that the hybrids (i. e., the heterozygote) for seed coat color "are often more spotted, and the spots sometimes coalesce into small bluish patches." In fact, Correns (1900) chides Mendel for failing to mention explicitly the detectability of the heterozygote, taking Mendel to task for overemphasizing the dominant-recessive relationship of the factors he used, and points out that the phenotype of the heterozygote for the factor for seed coat color was distinct from either homozygote.

Finally, it should be noted that Mendel did not give the actual details of the procedures used in this experiment, but he does state that "this experiment was made in precisely the same way as the preceding one." Since the preceding experiment involved seed characters only, without the necessity of making progeny tests, it is reasonable to assume that this trifactorial experiment also did not involve making progeny tests.

So it is that a critical examination of the most studied of all scientific papers leads to new questions whose probable answers give a new insight into the scientific methods of a century ago, and incidentally provide the examiner with the personal satisfaction of seeing in those papers information not obvious to generations of more casual readers.

Views of Research:
G. H. Hardy and Curt Stern

There are undoubtedly as many views of the nature and value of research as there are practitioners in the field, all of whom enjoy the "pleasure of finding things out." Many have the satisfaction of knowing that some addition has been made to the sum of human knowledge, or that their work will, directly or otherwise, improve the lot of humankind by alleviating suffering and pain.

But there is another aspect, a highly personal one, which I have described as the "joy of figuring things out." The difference between the joy and the pleasure depends on the extent of the intellectual contribution. We can take a simple example from astronomy. It is possible to detect the presence of a new planet in a solar system by comparing photographs of the same section of the sky taken at different times, and if a point of light appears to have moved, further investigation may indicate that it is revealing the presence of a new planet. On the other hand, it is possible that an observer may notice that the ordinarily regular paths of several known planets are deviating from their expected elliptical paths, and on the basis of this he may conclude that the deviation is caused by the gravitational pull of a previously undetected planet. Then he may calculate the probable position of the new planet as well as its mass and period of revolution around the sun. The first observer who notices the movement of a spot on a photograph will surely enjoy the pleasure of finding things out, even though the same observation could have been made by a trained chimpanzee, or, as is actually the case, by a machine. But the one who has predicted the presence of the new planet by virtue of his intelligence will experience the "joy of figuring things out."

In some areas it is possible to use the intelligence, limited though it be, to make a highly personal contribution, and this is one which would ordinarily not be appreciated by those looking in from the outside. Such workers are sometimes confronted with questions about the value of their work both to themselves and to society. One can recall Faraday's response to a question about the usefulness (or, better, uselessness) of the early work on the nature of electricity: "Of what use is a newborn child?" On occasion scientists feel compelled to defend their work and their personal convictions openly.

A classic example of such a defense is that of G. H. Hardy, a mathematician of considerable reputation and well known to geneticists as one of the originators of that fundamental theorem of population genetics known as the "Hardy-Weinberg Law." Higher mathematics, the "queen of the sciences" is particularly difficult to justify to the average citizen. For instance, how can the study of the properties of mathematical formulations in ten dimensions be important to any aspect of life on this planet? Hardy, in 1940, wrote in the conclusion of a small book entitled "A Mathematician's Apology":

"I have never done anything 'useful.' No discovery of mine has made, or is likely to make, directly or indirectly, for good or ill, the least difference to the amenity of the world. I have helped to train other mathematicians, but mathematicians of the same kind as myself, and their work has been, so far at any rate as I have helped them to it, as useless as my own. Judged by all practical standards the value of my mathematical life is nil; and outside mathematics it is trivial anyhow. I have just one chance of escaping a verdict of complete triviality, that I may be judged to have created something worth creating. And that I have created something is undeniable: the question is about its value.

The case for my life, then, or for that of anyone else who has been a mathematician in the same sense in which I have been one, is this: that I have added something to knowledge, and helped others to add more; and that these somethings have a value which differs in degree only, and not in kind, from that of the creations of the great mathematicians, or of any of the other artists, great or small, who have left some kind of memorial behind them."

Curt Stern (1968) expressed his view in the following way: "The scientific career has been called a carnivorous god. Perhaps more appropriately it may at times appear a soul-devouring god. By no means however does it need to take on this aspect. Whatever dangers personal weaknesses and social pressures may present to the investigator, he can rise above them. He can retain the enthusiasm of youth which led him to contemplate the mysteries of the universe. He can remain grateful for the extraordinary privilege of participating in their exploration. He can incessantly find delight in the discoveries made by other men, those of the past and those of his own times. And he can learn the difficult lesson that the journey itself and not only the great conquest is a fulfillment of human life."

Certainly the role of the individual and his feeling of personal accomplishment are diminished by the increasing complexity of research, the necessity for expensive equipment (and the concomitant funding from private or governmental sources), the emergence of large groups dedicated to a single task, the requirement imposed from above that the research be at "the cutting edge" and the vastly improved means of communication submerging individual effort into one great enterprise. Still, we can imagine that there probably exists more than one solitary worker, sitting in his ivory tower in the outback of scientific civilization, armed with sealing wax and string, who is still concerned with the intellectual pursuit of the elusive and who, as a consequence, will on occasion have reason to shout "Eureka!" For him, as Stern has put it, the journey itself is a fulfillment of human life.

APPENDICES

APPENDIX A

The Fruit Fly

Drosophila melanogaster, commonly called "the fruit fly," is well known to anyone who has had an unprotected bowl of ripe fruit on the table during the warmer days of summer and fall. Quite harmless, it is not to be confused with the Mediterranean fruit fly which is a serious pest.

Drosophila melanogaster has many features that make it ideal for laboratory studies. In a period of less than two weeks, a male and female can produce hundreds of offspring within the confines of a half pint milk bottle, with only a ripening banana to serve as food. Using a simple hand lens, many features of the fly's anatomy are evident and some, like changes in eye color (white eyes) or wing position (as in heldout) are obvious without magnification. Microscopic examination of the chromosomes is easy and extremely instructive. The larval stage carries within the "salivary glands" chromosomes of gigantic proportions making it possible to see detail not found anywhere else in the animal kingdom.

APPENDIX B

Salivary Gland Chromosomes

One of the great features of Drosophila from the research point of view is the existence of the giant chromosomes of the so-called salivary glands found in the larval stage. In some species, as in D. robusta, these are so huge that they can be seen easily simply by holding a microscope slide carrying them up to the light. When examined closely each chromosome appears to be banded, with the bands varying one from the other by their density, spacing, and general appearance so that each chromosome segment is readily identifiable by its individual characteristics. However, their suitability for research not only varies considerably from one species to another but within a species their clarity is highly dependent upon the particular strain used, the culture conditions under which the larvae are grown, the temperature and other variables still unknown.

Ordinarily each individual carries two sets of chromosomes, one from each parent. The corresponding chromosomes from each parent are called homologs; in the Drosophila salivary chromosomes these homologs come to lie side by side and fuse, and, in the usual case, fuse so completely that the two appear to be a single unit. If, however, the chromosome coming from one parent carries some very minor and ordinarily inconspicuous changes compared to that coming from the other, when they fuse the two will be out of register at that point and the insignificant changes will be immediately obvious.

Clearly this becomes a gold mine for the worker who is working with two different species that will mate and produce larvae for examination, for those changes that have occurred since the two diverged from a common ancestor can be clear and unambiguous. A minor complication arises in many cases of interspecific hybridization in Drosophila, however; the homologs often do not fuse but simply coil around each other and in many cases will not fuse or even pair with each other but will stay single and unpaired for all or part of their lengths.

APPENDIX C

Selection Against Heterozygotes

To understand why the survival of some abnormal chromosomes is an unusual event, we can bypass the details of the biology of those abnormalities and draw a parallel with a more understandable situations. Let us suppose that we have a container of positively charged particles and another of negatively charged. Each may be stable individually, but if we add a portion of one into the other, the positive and negative charges combine to destroy each other in a flash of light, removing equal numbers of each. Clearly the survivors will be represented by the one present in greater numbers. When a parallel situation is found genetically, this phenomenon is called "selection against heterozygotes." It is found in heterozygotes for inversions within a chromosome and translocations of segments between chromosomes. In humans it is exemplified by the Rh factors, where children exhibiting *erythroblatosis foetalis* are heterozygotes, having received the Rh+ factor from the father and the Rh- factor from the mother.

APPENDIX D

Population Cages

One morning in 1939 Dwight Miller told me about a paper he had just read by two French workers, L'Héritier and Teissier, who had put flies with different mutants into a large box supplied with plenty of food, and allowed them to breed in competition for several generations. They showed that one type outbred the others and became the predominant form.

Miller found this fascinating because he had previously found an interesting chromosome rearrangement occurring in the wild in Drosophila algonquin, one carrying a unique inversion which, in theory at least, should be at a disproportionate disadvantage against any other more prevalent chromosome type. Wouldn't it be a smart idea, asked Miller, to use the same system to test in the laboratory the theory that his special chromosome would be at a disadvantage? I agreed that this sounded like a great idea, and we went together to try it out on someone more knowledgeable. Sturtevant was not available, but Dobzhansky was. After explaining the idea to Dobzhansky, Miller was chastened to receive a highly animated tongue lashing from Dobzhansky who said that it was foolish indeed to imagine that any lab setup could duplicate conditions in the wild, and that such an experiment would be meaningless. We left his lab with our enthusiasm considerably dampened and the idea was never mentioned again.

Interestingly, Dobzhansky, after moving to Columbia University, far removed from the mountains, deserts and canyons of the West, which had provided so much material for his work, embarked on population cage experiments himself. On p. 495 of the Oral History Memoir: "So it was necessary to invent an experimental technique to demonstrate that interpretation [of high selection values]. So, in New York, I started to build so-called population cages, following a recipe proposed in the 1930s by two Frenchmen, L'Héritier and Teissier, who built such a population cage for an entirely different purpose, working with mutants." It is not immediately clear what is meant by "an entirely different purpose," since both were meant simply to observe the change in frequency over time as two or more genetically different types competed for survival in a grossly overpopulated artificial environment.

APPENDIX E

Translocations in Other Species

Interchromosomal changes do not always have serious deleterious effects in producing a large proportion of inviable zygotes. In Oenothera, the evening primrose, a system for chromosome separation in translocation heterozygotes has evolved that ensures that balanced chromosome complements of translocation heterozygotes regularly separate, with relatively few errors, although it should be noted that almost all of the "mutations" described by Hugo de Vries at the beginning of the twentieth century were in fact viable products, with visible effects, of the occasional irregular unbalanced products from translocation heterozygotes.

In animals, rearrangements that involve whole chromosome arms are largely immune from the deleterious effects found in translocation heterozygotes because the pairing of homologous segments is continuous in the active euchromatic regions and separation of homologous segments can proceed normally. This absence of, or reduction in, selection against chromosome arms when separated, or attached, has been inferred in Drosophila since the pioneering work of Metz and Moses (1921), and is nicely illustrated by the chromosome variants in D. virilis, where several chromosome arm variants are known and appear to be innocuous.

It is for this reason that the stable units, called elements, have been pinpointed as the chromosome arms rather than the chromosomes themselves (Muller 1940, Sturtevant and Novitski 1942b). It should be emphasized however that the wide variation in chromosome number and form of various species within the genus Drosophila constitutes strong evidence that gross aberrant changes may be incorporated during speciation and suggests that there may exist factors (such as meiotic drive) beyond those now considered important that promote their incorporation. It is for this reason that the appearance of apparent translocation differences in closely related species deserves close examination.

Whole arm rearrangements found in humans are often called "Robertsonian translocations," although they are not translocations in the usual sense of being exchanges of substantial amounts of genetically active material between two unrelated chromosomes. They are often found because the occasional unbalanced gametes involving small amounts of genetic material end up in inviable fetuses or abnormal newborns.

A particularly interesting whole arm rearrangement, not seen in any other form cytologically studied, is found in humans in which two independent chromosomes present in all other hominids have become attached at their tips, in the telomeric regions, and have formed chromosome 2 in humans. Once again, it seems reasonable to assume that at the time of the attachment, because the pairing of two independent arms could proceed without difficulty with the two newly attached arms and produce (entirely, or almost entirely) balanced germ cells, the heterozygote would not suffer from the pronounced negative selection characteristic of typical translocations.

It should be noted that, in addition, nine cases of inversion differences, intrachromosomal rearrangements, have been identified as differentiating the two hominids, humans and the chimpanzee, and that every cytologically identifiable chromosome segment present in man has been found on the corresponding chromosome in the chimpanzee. Furthermore, assuming that chromosome breakages leading to rearrangements occur at random among the chromosome arms, the nine intrachromosomal differences between human and chimpanzee would suggest that perhaps as many as 180 reciprocal translocations should have originated during the same period of time. That there are none now suggests that as fast as they appeared they were eliminated. This is a strong argument that the concept of the integrity of chromosome elements as evidenced by the similarities and differences between closely related species is not limited to Drosophila, but may be found as well in other closely related species such as human and the chimpanzee.

APPENDIX F

Adventures in Mathematics

Our foray into computer analysis was a pioneering effort, the first for us and, I believe, the first in the field of genetics. It was remarkably productive. We found (or, better, the computer found) a serious conceptual defect in the way that population geneticists were regularly analyzing their data (Novitski and Dempster 1958).

When two alleles (or two distinctive chromosomes) are found in a population and a count is made of the frequencies of the three combinations, the two homozygotes and the heterozygote, it is more often than not observed that the heterozygote appears to be in excess. Population geneticists ordinarily attributed this to "heterozygote superiority" and this concept has played a significant role in the formulation of their ideas of the nature of natural selection.

A graduate student (William Hexter) at Berkeley had, for his Ph.D. thesis, put various recessive mutants of D. melanogaster into population cages along with normal flies, and recorded over many months the gradual decline in the frequency of the mutant flies. Since the recessive gene was found not only in homozygotes, where it could be seen and counted, but in heterozygotes, where it could not, the calculation of the overall frequency of the recessive allele depended upon the algebraic estimation of the number of heterozygotes carrying the gene. In collaboration with Everett Dempster, an electrical engineer turned geneticist who had served on Hexter's Ph. D. committee, we decided to see if we could program the ORACLE to give us the best estimates of the relative viabilities of the various gene combinations as they changed over a series of generations.

The amount of data was considerable and the resulting algebraic formulations were cumbersome and extensive, not amenable to solution by ordinary means. The chance of success seemed good, provided the program we envisioned could be made to work properly. Its task was to evaluate a hundred or so observations from different experiments and to come up with the best estimate of the relative viabilities of the three classes. This it should be able to do with ease once we supplied it with initial arbitrary guesses as to what they might be. However, the computer came out with different values for our arbitrarily assigned starting points and we then realized that the reason was that our data, voluminous as they were, were fundamentally flawed. There was a built-in ambiguity because we had more unknowns than we had equations tying them together.

Dempster and I realized that this ambiguity was to be found also in the simplest cases and unnoticed in just about all of the publications resulting from population work being done at that time. At a meeting of the Genetics Society, I discussed this with M. J. D. White, an eminent cytologist who had published several excellent books and many papers, some of which had this fallacy. He did not appear to understand; his linguistic and naturalistic capabilities clearly exceeded his mathematics. However, when I discussed this with Bruce Wallace, one of Dobzhansky's several mathematically oriented students, his reaction was of immediate comprehension, In fact he quickly wrote a simplified and more understandable version of our argument and submitted it to the journal Evolution which, apparently lacking a lengthy review process, published his paper before ours came out in the journal Genetics. His note (Wallace 1958) with two simple arithmetical illustrations, is meant for the linguistically oriented; a more elegant mathematically rigorous derivation suggested by Prof. James Crow is presented in endnote 22-1.

ENDNOTES

(1-1)   For one thing, both Sturtevant and I had unusually short undergraduate careers. He had finished his degree at Columbia in two years and I did the same at Purdue. Also each had published his first scientific paper while an undergraduate.

(3-1)   This experience involved manual typesetting by hand, which necessitated reading type upside down and even backwards, a skill which never left me. I also learned the origin of the nonsense phrase "etaoin shrdlu" which would unaccountably appear at random in newspaper columns of those days. Back in those days when type was set on a Merganthaler linotype machine, the operator would, upon detecting a typing error, simply run his finger along the row of characters on the top row of the keyboard. This would make the error obvious, and the line could be easily identified, discarded and retyped. The appearance of the nonsense phrase in the final article simple represented cases where the deliberately ruined slug of type was accidentally left in place. I was so enthralled with the linotype machine that I decided that there was no more noble profession for me in the future than to become a linotype operator.

(3-2)   Miss Tyburski seemed to treat me more as a colleague than a lowly student. I suspect that this stemmed from an interest in promoting the welfare of a student with an ethnic background similar to hers, one who might therefore be at a disadvantage compared to others in the class. At one point she showed me the "IQ" scores for all the members of the class. For what it is worth, and most psychologists these days would say "not very much" the person who scored highest was an otherwise somewhat inconspicuous student named Sylvia Guzy who reached the good score of 160. I was second, somewhat lower, pretty good, considering that I was completely deficient in that aspect of intelligence designated as "spatial intelligence" that occupied a good fraction of the test. Eddie Lewis scored at 120 and when I suggested that perhaps this was too low, Miss Tyburski assured me that this was an accurate measure of his abilities. Considering his later considerable success in his field, it suggests that there may be factors leading to success in science far more important than that represented by the questionable IQ number.

(3-3)   I was almost given an opportunity to give a speech in another connection. The teachers decided that it would be interesting and profitable to the students to give them the opportunity to give a short talk at the regular morning convocations which all students attended. This I refused to do. However, my good friend Eddie Lewis, being made of sterner stuff, volunteered. When, several days later, he got up to deliver his prepared talk, he promptly slumped to the floor unconscious. (Presumably pressure on the carotid artery cut off the blood supply to his brain.)   In any case, that session was abruptly terminated, much to the delight of the students, and the program itself was cancelled, undoubtedly because in future talks the attention of the students would be diverted from the substance of what the novice orators would have to say to the question of how long he or she would remain standing!

This dread of appearing before an audience never left me. Even in later years when I regularly appeared before classes of a hundred or more, I felt distinctly ill at ease, although, I hope, this was not obvious to my audiences. I was surprised to learn subsequently that this discomfort was shared by many others, some of whom were known as accomplished speakers (Curt Stern, for instance).

(4-1)   A quick search on the internet suggests that more than 500 research papers have been written concerning this locus over the past several decades.

(5-1)   It was later explained to me that the reason for this question was that Columbia University and the other New York universities had a disproportionate number of very bright and capable Jewish graduate students who, once having gotten their degrees, could not find academic jobs in other universities because of the pervasive anti-Semitism found in academia throughout the country. For this reason a promising applicant, non-Jewish, would be most welcome.

(5-2)   A few days after Bridges died, I was asked by Jack Schultz to gather together Bridges' stuff from his lab and to dispose of it. This was at first fairly simple, someone had apparently preceded me and had taken care of various items. What was left was disposable. Odds and ends of junk went into the garbage.

But then I came upon some of the original drawings of Drosophila mutants made by Edith Wallace, the illustrator for the Drosophila group, for the classic "Dutch Book (1925)." I thought that it would be criminal to throw them out, and I saved them. They were stored away during World War II and for the next fifteen years they were distributed one by one to Ph.D. students. I imagine that some of them are still hanging on the walls of universities around the country, being glanced at occasionally by passing students who could have no idea of their interesting history. They can be identified because they carry the name of Edith Wallace, pinx, the word "pinx" being the abbreviation for the Latin pinxit, meaning he/she painted it.

I also found some unprinted negatives in a special wooden holder that Bridges had fashioned. This I saved too, and these were printed some twenty five years later.

(6-1)   Koller was an exceptionally humorous and good-natured fellow except for one unfortunate situation about which he continually complained early in the summer. It seems that Koller's room would become unbearably hot and he would suffer the tortures of hell, a punishment he felt was visited upon him by the Almighty for his having forsaken his vows. After more than a week of this misery, the problem was looked into by the then director, Milislav Demerec, who, discounting divine intervention, solved the problem. It seems that Koller's room, which was next to the mess hall kitchen, had a large connecting ventilating conduit that had been inadvertently left open and the heat from the kitchen stoves was pouring into Koller's room.

(6-2)   I subsequently learned that about half a dozen others had also pointed out the problem with the backwards moving chromosomes, and that each claimed credit for being the first to point it out. Probably I was not the first.

(6-3)   Commenting on Sturtevant's lack of interest in pursuing this work, Dobzhansky says on p. 411: "Thinking of this matter at present, I do not understand just why he started this work and dropped it. It seemed to me then, and continues to seem now, that this was a most interesting lead. There was a man who started this work; possibly he somehow did not like to think that what he was doing was merely confirming the findings of somebody else. He did not think much of either Chetverikov or Dubinin." Dobzhansky rather clearly reveals his confusion about the differences between himself and Sturtevant in their approaches to scientific problems, and finishes his comments with a non sequitur about two Russian workers.

(6-4)   Again, on p.404: "I started to do it with geographic strains, in a way which Sturtevant was not interested, in fact did not approve." This is the point alluded to earlier, that Sturtevant had essentially disapproved of the general trend of the thesis problem Dobzhansky suggested that I work on, long before that curious encounter in the basement of Kerckhoff. Only after reading this in Dobzhansky's Oral History Memoir did I have any idea why Sturtevant was so noncommittal and was hurling the chunks of ice against the wall with unusual energy.

(7-1)   More detailed information on Sturtevant's career may be found in Lewis (1995), Crow (1987) and Provine (1971).

(7-2)   During those early days, it was not uncommon for friends and relatives to express wonderment that anyone would find any reason to spend time studying the fruit fly.

(7-3)   Excellent accounts of the early history of biology at the California Institute of Technology as well as the Drosophila group will be found in Goodstein (1991). See also Carlson (1966), and Kohler (1994).

(7-4)   Dobzhansky never received a Ph.D. degree, which is the standard entrance certificate for an advanced academic position in a science department at most universities. It is possible that this gap in his training may have to some extent been responsible for his unique approach to scientific investigation, a point which is discussed by R. Lewontin, in a later chapter. It was my impression that he was quietly amused when graduate students addressed him as "Dr. Dobzhansky," which they usually did.

(7-5)   At several points in his Oral History Memoir Dobzhansky notes that it was Sturtevant who was his main proponent during those first days. Thus, on p. 269, "It is unquestionably to Sturtevant that I owe my having been invited to stay in this country. Consequently it is to Sturtevant that I definitely owe my life, since without doubt I would no longer be alive if I had gone back to Russia."

(7-6)   Unlike just about everyone else, Sturtevant did not have a high power microscope in his lab-office during the period from 1938 to 1942, or any time thereafter I should think. During the period of two years that I was his "research assistant" he would occasionally ask me to look at the chromosomes of some species or other. It would take me several hours to make a preparation and find an unambiguous configuration that might give him the information he wanted. When I would call him to my lab to verify the chromosome configuration, he would glance at it for a few seconds, grunt his OK and walk away.

(7-7)   All in all, Dobzhansky's publications amounted to more than forty during the period from 1927 to 1936, a formidable total for a biologist. (Sturtevant's total during that same period came to less than fifteen.) After collecting data (with assistance whenever possible) he would always publish, keeping count of his publications and years later referring to papers by the number in the order he kept of his publication list. At least, this is my impression. When I visited him at Columbia University in 1943 and I asked about a particular paper, he said "Oh yes, that was number 37" (or some such number).

(8-1)   The problems that they posed in that book (ostensibly for elementary students but actually useful only at a much higher level) were so difficult that they themselves sometimes gave incorrect answers. Sturtevant was away during the 1939 academic year as a visiting professor at Harvard. When he returned he reported with a mixture of chagrin and delight that the graduate students there had a number of corrections to make of the answers that he and Beadle had provided for the questions in their text.

(8-2)   This was a point that they must have realized, but preferred not to make an issue of because it was irrelevant to their argument. The reason for this will be discussed in a subsequent section.

(8-3)   It might be noted that at least one of the inversion paths used in the phylogeny involved was not overlapping, but one was an included inversion, a point which Dobzhansky undoubtedly realized, but which was not mentioned because it did not alter the overall argument appreciably (Novitski 1946).

(10-1)   These are not my only inborn deficiencies. If one adopts some modern psychologists' division of mental capacities into from seven to ten more or less independent facilities, I am severely limited in a "naturalist's" intelligence, that capability which leads a person to compare minute differences in similar items, and to classify and name them. Early in my education, I had serious doubts about going into biology as a career because of this impediment. It is this facility that can make a person an outstanding biologist when that same person has average capabilities (or less) in other aspects of intelligence (Charles Darwin has been mentioned in this connection earlier). It is also an aspect of intelligence that is downgraded by those in other "hard" sciences (mathematics, physics, chemistry) where the logical-mathematical aspect of intelligence is not only necessary but is considered the only true type of intelligence!

Another deficiency of mine is in the area of spatial relationships. I became aware of this when in high school I took my first "intelligence" test. I score pretty high on such tests because I know that such tests depend on a maximum number of correct answers in a given subject, speedily given, in a limited period of time. I do very well except for that part of the test that has to do with shapes, views and orientation of geometric figures. Here, after answering the first couple of simple-minded questions, I am completely at sea and have to guess at practically the whole section. In recent years, my deficiency in spatial perception has become more noticeable in a more down to earth setting. After having lived in the same house for more than forty years, I find it impossible to relate the floor plan of the main floor to that of the basement or to that of the second floor. I also have a defect of vision, red-green color deficiency. This has been discussed in more detail elsewhere.

(10-2) One incident I recall gave me a lesson in human fallibility. During my first Christmas season in Pasadena, I did what all newcomers did: I watched the Rose parade. As I walked along the projected route, I had no difficulty finding a good viewing spot. In fact there were large vacant stretches available, unoccupied by spectators. After a short time watching the floats roll by, I got bored and left. The next day the Pasadena Star News reported that several million people had lined the streets to watch the parade. At tea time the next day I commented that I had made a quick calculation: if one took the length of the route and the space occupied by the width of the average viewer, the spectators would have to been about eight deep along the entire route to give the newspaper estimated number. The newspaper estimate of several million was clearly an exaggeration. My calculation appeared to have the approval of all present, a group that included Albert Tyler, the distinguished embryologist. I forgot about this, until a few days later there appeared in the Pasadena Star News a letter to the editor with the exact calculation I had made at tea time some days earlier. It was signed by Albert Tyler. Apparently he was half asleep when I had explained this, or he had considered the source of the idea irrelevant to the point he wished to make. In any case, this was my first exposure to the fact that there is no personal possession of ideas by scientists, nor, for that matter, any necessity for acknowledgment that the essential point had first been advanced as a communication from another. This is not uncommon, and every active scientist will at one time or other find that his "original" ideas are adopted by others, and elaborated upon, without any attribution. This may be considered the accepted norm by the non-scientist, but it can be a bone of contention by many if not most scientists. Among the most vocal about the necessity of appropriate acknowledgment was H. J. Muller, who was probably the most prolific idea generator in the genetics of his time (11-3).

(11-1)   During my first year, Dobzhansky reported to me one morning that he and Natasha had the previous night been to a concert at the Hollywood Bowl where they played Tschaikovsky's seventh symphony. Without thinking, I blurted out that I thought that that composer had composed only six symphonies (it did not occur to me to ask whether he might have meant instead Beethoven's Seventh!). Dobzhansky's face darkened and he refused to speak to me for several days thereafter. Why he did not instead laugh and admit to a simple slip of the tongue is beyond me.

(11-2)   Despite this disagreement on research problems, my relations with Dobzhansky went along reasonably smoothly. But when Esther and I married a few years later and I spent a few days in New York, awaiting the arrival of the Queen Elizabeth to take me to the European Theatre as a radar officer, we had no qualms about visiting Dobzhansky at Columbia University. He was very polite, showed us the rooms where the original fly lab of the Morgan group was located (then used as a store room) and where his lab was located. However, beneath the surface was an unmistakable coldness that made me uncomfortable, and I said my adieus sooner than I had anticipated.

(11-3) In a review of the book on Muller by E. A. Carlson, I wrote: "Muller's keen and prolific mind digested the elements of a problem quickly and, in rapid succession, suggested possible solutions. Obviously this ability is welcomed by any group engaged in experimental work. But when someone engaging in such speculation over an extended period of time demands priority recognition for those suggestions that were borne out by the experimental work of others (and forgets about those that were not) he inadvertently relegates his colleagues to the awkward position of serving as his technical assistants and is likely to become an unwelcome guest in that laboratory. Muller had the misfortune of having a hyperactive mind and a predilection for being right more often than not."

(11-4) About three decades later I visited Caltech as a member of a National Institutes of Health committee to evaluate their biology graduate program when they were applying for a continuation of a training grant which provided them with several hundred thousand dollars per year. It turned out that the faculty member in charge of the program was Prof. Bonner. Our initial contact was, for a few seconds, slightly strained; I think that the irony of the situation, the reversal of positions as examiner and examinee, was evident to both of us. However, all members of the committee eventually agreed that the program was excellent, the only criticism being that the graduate students were evidently so happy in the magnificent Southern California climate that they were taking an unusually long time to get their Ph. D. degrees!

(12-1) It may seem difficult to understand how they could have made this oversight of overlooking a chromosome arm. They drew by hand the giant salivary gland chromosomes in minute detail, with several hundred delicate transverse bands, so extensive that it required a folded insert in the journal to reproduce their drawings. But Socolov was working only for the summer, while away from his university in Mexico; perhaps they felt pressure to finish the manuscript before he left.

(12-2)   The interest in this problem goes back to the 1920s, when two workers, C. W. Metz and M. S. Moses (1923) made a study of the chromosome configurations in a wide variety of different species of Drosophila and found that they fell into definite patterns. It was not that the configurations themselves were identical, but that the total number of large chromosome arms seemed to center on five. Muller referred to these units as elements. Both Sturtevant and Muller realized that the reason for this relative uniformity rested in the elimination from the population of those chromosome changes that would mix pieces of the chromosomes together, or subdivide them into smaller pieces, as translocations would. Sturtevant was trying to find evidence that would support (or refute) the idea of the relative constancy of the elements by collecting information on the mutant genes found in as many species as possible, together with their relationships to each other.

(13-1)   Morgan's relationship with Dobzhansky seems to have been more favorable than Sturtevant's. It seems clear to me that this resulted from a coincidence of their intelligences, that Morgan, like Dobzhansky, was primarily a naturalist, with a strong linguistic component. In effect, they talked the same language.

Morgan's deficiencies in the math area became evident to me during my first few months as a graduate student. He gave a seminar that autumn on his studies on the self-sterility alleles in the sea squirt, Ciona. Towards the end, he made some simple calculations on the probable total number of such alleles using what is known as the Poisson distribution. Sitting in the back of the room with Dobzhansky, I whispered that Morgan obviously was on precarious grounds, since that procedure demanded a random distribution of alleles, and there was every reason to believe that in this instance this was not so. Dobzhansky told me to keep quiet, obviously in deference to the old father figure, who was giving what I think was his swan song. When that account was published some time later, the offending math had been removed.

(14-1) Direct examination of the progeny of freshly caught fertilized females has revealed occasional hybridization. For detailed discussion see Noor (2000).

(14-2) Drosophila novitskii, for instance, got its name because Dwight Miller found the original specimens near our outhouse in Gothic, Colorado!

(15-1) During the years of 1938-1939, when we went on all of his collecting trips, not once were his daughter or any other relatives included. His daughter's trips must have occurred either before 1938 or during the few months before he left for Columbia.

(17-1) This change in attitude may have, in part, been dictated by the limitations on the types of "experiments" available in New York City, when his primary interest centered on a species found exclusively in the western United States.

(17-2) It is worth noting that the three previously named workers, Streisinger, Pittendrigh and Zimmering, worked in the Dobzhansky lab for only a short period of time, all moving into fields where the strict experimental method predominated, and all doing unusually creditable work. It can be assumed that one or more found the Dobzhansky style of research not compatible with their mind-sets, just as I did much earlier.

(17-3) One other theoretical possibility is that Dobzhansky started out with more than four lines, and simply dismissed the data from the contaminated line. However, his description of the experimental design in the introduction of that paper seems to leave no room for that possibility.

(20-1) I have sometimes been categorized as being a student of Sturtevant's because both of us showed interest in the same area of basic biology. In the strict pedagogical sense this is misleading. It is true that as a student I worked on the same floor of the same building as he did, and had frequent conversations with him, but these were never in the area for which he was so well known, and when he suggested a thesis problem and approved it with his signature, he was doing so in the area of population genetics, an area quite removed from the one now under discussion, that of chromosome behavior.

From time to time persons removed from the actual workings of the scientific enterprise construct "genealogies" showing the transmission of scientific thought from one set of workers to their descendants. The lines of descent are based sometimes on the person who signed their Ph. D. thesis, or on the laboratory in which they happened to do their work, or on the general subject of their research interest, either as a doctoral thesis or in their later careers. Sometimes this is confused by the student's preference that his name be associated with a well-known worker or, less often, that he prefer for personal reasons not to acknowledge the contribution of his mentor whom he considers to be less well-endowed intellectually. In any case, these scientific pedigrees usually end up as a hodgepodge of all the factors given above, with no possibility of discriminating among them; the selection of the criteria is determined arbitrarily and randomly by the historian of science.

I myself would prefer that the basis for such genealogies, if they are to exist at all, center on the inheritance of the conceptual framework of the research. From this point of view, then, it might be appropriate to say that I was a student of Sturtevant. When there has been no actual personal contact of the individuals involved it becomes less obvious. A case in point is that of Ed Lewis, who arrived at Caltech with a research project involving the mutants Star and Star-recessive and who worked on this project first at the University of Minnesota and then at Caltech for several months before he mustered the courage to ask Sturtevant whether he could proceed with it as a possible Ph. D. thesis.

However, it was Calvin Bridges who steadfastly maintained that the detailed study of individual loci in Drosophila would have far-reaching consequences, not only for the elucidation of gene structure, but also for the understanding of the role of the gene in development. He was the first to make detailed maps of the chromosomes of D. melanogaster, both genetically and cytologically, and he tested alleles with subtly different morphological manifestations to see if they were actually at the same locus. Using the criterion given above, I would say that Lewis was the student of Calvin Bridges (whom he had apparently never met), and not A. H. Sturtevant, whose involvement was minimal, although officially he was Lewis's "thesis advisor."

(21-1)   In the early 1960s I was surprised to read in Nature an article by Leo Szilard (of atomic bomb fame) in which he presented a possible reason for the change in human birth sex ratios as parental ages increased. Apparently in his effort to get away from nuclear physics, he had decided some years earlier to dabble in biology, and this was his latest interest.

However, he stated that because the ages of spouses were correlated, it was impossible, on the basis of available data, to correlate an age effect on the sex ratio at birth to either of the two parents. Of course I was aware that such a separation had been made earlier (Novitski 1954), and that the age of the father had been pinpointed in several subsequent publications as being most closely correlated with the changes in sex ratio as parents aged. This would be a matter of some interest to Szilard.

Wanting to discuss this matter with Szilard, I inquired as to his whereabouts from his good friend, scientific collaborator and my university colleague, Aaron Novick, and learned that he was in Memorial Hospital in New York City, being treated for terminal cancer. Under the circumstances it seemed to me probably inappropriate to discuss scientific matters, but a quick check with the hospital assured me that not only would such a visit by me be acceptable, it would be welcomed.

A trip to New York took me to the Memorial Hospital where I found him quite alert, quite interested, and surprised to learn that in fact our work had eliminated the doubt he expressed in his paper about the statistical ambiguity. I gave him the latest paper we had put out on the subject (Novitski and Kimball 1958) and we had about half an hour of pleasant discussion on the topic. I do not recall that I discussed with him my own somewhat negative reaction to the fundamental assumptions in his proposal, that a cell initially carrying both X and Y chromosomes but eventually lacking a Y-chromosome after its destruction by radiation would proliferate normally, go through the male meiotic divisions, and differentiate into mature sperm, all carrying only the surviving X chromosome.

I left, satisfied that I had given something positive to mull over, surely not enough to obliterate the thoughts he must have had about the fate that awaited him in the next few weeks or months, but enough to occupy a few moments of his remaining time. For my part, I was gratified to have had the opportunity to meet with, and to discuss scientific problems with, one of the most remarkable scientists of the twentieth century.

(21-2)   One might think the distinction between a rod-shaped chromosome and a ring chromosome would be most clearly obvious by actually looking at the chromosomes microscopically. Not so, at least in this case. The terminal tuft of inert material at the end of the opened-out ring would cause the rod to bend back on itself and assume a deceptive ring-shaped configuration, whereas, in a genetic test where crossing over between the unknown chromosome and a normal one is measured, recovered crossovers involving a ring chromosome occur only one fourth as often as the complementary ones involving the rod, a clear and unambiguous indication that the chromosome of unknown structure is a ring.

(21-3)   The knowledgeable geneticist will recognize this method as one which is widely used in work on microorganisms where the culture medium can be manipulated to select automatically for genetically desirable types.

(22-1) An original mathematically rigorous proof that a depression of one of the two homozygous classes will lead to an apparent "superiority of heterozygotes" has been developed by Prof. James Crow and is presented here:

REDUCED VIABILITY OF ONE CLASS MIMICS HETEROSIS

Suppose that a population is in Hardy-Weinberg ratios after fertilization, but that one genotype has its viability reduced by an amount X. Then we have

Genotype	AA	AA$'$	A$'$A$'$	Total	
Zygote Frequency	A	2B	C	1	B^2=AC
After Selection	A/D =A$'$	2B/D=2B$'$	(C - X)/D= C$'$	1	D = 1-X

Half the expected proportion of adult heterozygotes:
$$A'C' = (AC - AX)/D^2$$

Half the observed proportion squared of heterozygotes:
$$B'^2 = B^2/D^2 = AC/D^2 > (AC - AX)/D^2$$

So, the observed proportion of heterozygotes is greater than the Hardy-Weinberg expectation.

(25-1) I put the question of translation to a native German, an architect (who would have no preconceptions as to preferred meaning), and he responded in the following way:

"etwas anbauen" has not very many meanings.
The one that immediately comes to my mind is archi
tectural, reflective, and means "to add something on," like in "adding a porch on to a house."
The obvious other one is agricultural and means "to
plant something",
although in the context below I would have chosen
"anpflanzen," or

"pflanzen", which is apparently of the same lingual
root as "plant."

"anbauen" usually means a bigger scale endeavor,
like "Mais anbauen"

(to plant or grow corn), at an economically feasible
scale.

So, as much as I can tell from the fragment below it
pretty much means:

"in order to test the meaning (or significance) of
that, planted 10 seeds of each."

REPRINTS OF

RELEVANT PAPERS

THEODOSIUS DOBZHANSKY—

A THEORETICIAN WITHOUT TOOLS

Richard C. Lewontin, Harvard University

The field of population genetics is unique in biology in the degree to which experimental and theoretical aspects are so tightly intertwined. The current fad for molecular population genetics, for example, is entirely motivated by the possibility of using molecular data to estimate the selective and population structure parameters of the synthetic stochastic theory of population genetics elaborated by the Wrightian enterprise. The intimate relation between mathematical theory and experimental and natural historical observation is a continuation of a mode of science that we owe in large part to the work of Dobzhansky. This mode, more than any specific problem or theory, is the real continuing significance and influence of his work.

Theodosius Dobzhansky is thought of as one of the founders of experimental population genetics and especially as the person who brought together laboratory experimentation and observations of the genetics and behavior of natural populations in an attempt to explain the dynamics of genetic variation in nature. In a field that spans the modes of scientific work from the abstract probabilistic theory of Wright and Kimura to the ecological field studies of Cain and Heed, Dobzhansky is always placed firmly at the experimental end. I argue in this paper for a very different view of Dobzhansky. I claim that he was, in fact, a theoretician whose entire intellectual program was theoretical and conceptual but that, lacking the abstract mathematical tools that are the stock-in-trade of the conventional theoretician, he used the only tool at his disposal, the manipulation of living organisms. He was, in this view, an experimentalist *faute de mieux*, for whom organisms were a kind of analogue computer, rather than the subjects of interest in themselves. I bolster this heterodox view by a sketchy analysis of Dobzhansky's empirical work and by a clear look at his theory of *coadaptation*.

To anyone who was familiar on a day-to-day basis with Dobzhansky, the claim that he was a theoretician can seem only perverse. He repeatedly told his students that "science is 99% perspiration and only 1% inspiration" and that "statistics is a way of making bad data look good." He had a repertoire of stories of the follies of statisticians and theoreticians, such as the claim that C. I. Bliss once definitively misidentified a Live Oak in Pasadena as an orange tree on the grounds that 99.5% of all trees in California are orange trees. Dobzhansky himself was innumerate, or at least he appeared to be. He could not carry out and claimed not to understand the simplest algebraic manipulations and was unable to use the tables of the standardized normal deviate to draw a normal curve with a known mean and statistical deviation. He would on occasion demand from a theoretical colleague a formula that could be put to some use in manipulating data, but he never asked how the formula was arrived at.

This picture of Dobzhansky, familiar to his students and postdoctoral fellows, comes as a surprise to those who know his work only from publications. No other large corpus of empirical biological work is as deeply dependent upon and intertwined with mathematical theory as is Dobzhansky's "Genetics of Natural Populations" (GNP) series (see Lewontin et al. 1981). These papers use a variety of theoretical formulations of Sewall Wright and employ a number of techniques of estimation of the parameters of Wright's models, including sophisticated simultaneous least squares and regression methods. Moreover, the papers are characterized by considerable statistical analysis, including nonparameter methods and analysis of variance at a time when most biologists, even evolutionary biologists, were strangers to statistics. It is not surprising that the body of Dobzhansky's work on population genetics would give the reader the impression of someone very familiar with and at home with the most advanced theoretical biology of the time.

A closer examination of the papers and a knowledge of Dobzhansky's actual practice reveal the actual situation. Dobzhansky collaborated constantly with theoreticians. Four of the GNP series were coauthored with Sewall Wright; two with H. Levene, who also appears as coauthor on three other papers not in the series; and Dobzhansky's theoretically sophisticated graduate student and colleague Wyatt Anderson is coauthor on five of the GNP series and on two papers outside the series and is sole author on GNP 41. Beyond formal coauthorship, virtually every paper in the GNP series acknowledges the theoretical and statistical advice and work of W. Anderson, H. Levene, R. C. Lewontin, S. Wright, and B. Wallace, all of whom designed and carried out statistical tests, wrote computer programs to estimate parameters, and even designed the experimental protocols according to theoretical and statistical principles.

What is most important, however, is not Dobzhansky's dependence upon statisticians and theoreticians for the mechanical details of data collection and analysis but the central role that theory played in the formulation of Dobzhansky's problematic, not only in general but also in specific papers. An excellent example of the hidden hand of theory at both levels is in GNP 6, "Microgeographic races in *Linanthus parryae* (Epling and Dobzhansky 1942). Ostensibly a collaboration between an experimental genetical evolutionist and an evolutionary botanist to describe the variation in flower color along a geographical transect, the intellectual program of the paper arose directly out of Wright's theory of stationary distribution of gene frequencies under random genetic drift. Wright is acknowledged for help in the statistical analysis, but the real importance of Wright can be seen in figure 4 of the paper, showing "the frequency of samples containing different proportions of blue-flowered plants." Such a frequency distribution makes no sense at all outside the context of Wright's stationary distribution theory. It is precisely the *negation* of the then unusual treatment of such transect data that would have sought an ecotone or cline related to the external environment. Indeed, it is unclear why anyone would even want to make the observations in the absence of Wright's system of explanation. Note that Wright quite independently undertook a reanalysis of these data as part of his own program (Wright 1943, 1978).

The paper on *Linanthus parryae* leads us to the most general consideration of the structure of Dobzhansky's intellectual program. The functions of empirical work in biology are diverse, quite aside from those observations meant simply to augment the catalogue of diversity of living and dead organisms. A major mode for analytical and reductionist biology, characteristic of nearly all physiological, molecular, and developmental biology, is the elucidation of the material mechanisms that mediate biological phenomena, the description of the gears and levers of their articulations, such as, for example, the determination of the structure of DNA by Watson and Crick or the determination of the DNA decoding mechanisms by Khorana, Brenner, and Crick.

At the other end of the scale of description are experiments meant simply to rough out the description of the phenomena themselves: what happens if one sees the world through inverting prisms, inserts a bit of dorsal lip of a blastopore into the hind end of a gastrula, removes all members of a competing species from a community, and so forth? Then there are the classic experimental tests of hypotheses, a great deal rarer than older textbooks on the philosophy of science lead us to believe, but occasionally appearing in their idealized form, like Meselson and Stahl's test of the semiconservative model of DNA replication by isotope-labeled precursors. Dobzhansky's work belongs to none of these modes. Rather, it is of two sorts: the experimental measurement of parameters of an already articulated theory and the illustrative demonstration of what were, for Dobzhansky, unquestioned general truths about nature through concrete examples.

One of the remarkable features of Dobzhansky's empirical work was that although he spent a large part of his career working on the selective importance of new mutations and inversion polymorphisms, he devoted no attention to the elucidation of the ecological and physiological variables that underlay the balanced polymorphisms in which he fervently believed! With the exception of an aborted attempt later in his experimental career to link changes in the frequency of chromosomal inversion in *Drosophila pseudoobscura* to insecticide use (Anderson et al. 1968, GNP 39), and the earlier suggestion (with no test of the hypothesis) that several years of drought might have been the cause of inversion frequency changes in *Drosophila pseudoobscura* and *D. persimilis* over a four-year period, Dobzhansky evinced no interest whatsoever, at least in his active scientific work, in finding out what ecological and physiological factors were responsible for the fitness differences inferred in nature. Note that the paper on insecticides is the only one of the GNP series with the word *test* in the title. When general principles are involved, the papers are never characterized as "tests" but sometimes as "proofs." The problem for Dobzhansky was to demonstrate selection in action, not to elucidate its mechanisms. In this sense Dobzhansky's work on selection was purely formal and curiously unbiological for someone who thought of himself as a naturalist who spent much of his life outdoors and who enjoyed an immersion in the diversity of nature.

The formality of Dobzhansky's work on the inversion polymorphisms is seen in the two paths he took to demonstrating selection. Having shown that the repeated cycles of inversion frequency change in California could not be explained by migration of flies from one altitude to another as the seasons changed, his first demonstration that selection was occurring was to show that selection would change inversion frequencies in the laboratory. These experiments (Wright and Dobzhansky 1946, GNP 12), although entitled "Experimental *reproduction* of some of the changes *caused by natural selection* in certain populations of *Drosophila pseudoobscura*" (my emphasis), *reproduced* nothing. They were not intended to actually mimic the natural ecological situation, a hopeless goal in any case, or to isolate the causes of the selective differences, or to reproduce the seasonal cyclicity or characteristic frequencies of any of the inversions in any natural populations. Rather, they demonstrated that it was possible, in a highly crowded laboratory population, at a constant temperature of 25°C, neither of which conditions was "natural," to observe repeatable changes in inversion frequencies and eventual stable equilibrium of intermediate frequencies and to demonstrate that these changes were the result of differential fitness of genotypes with heterozygotes having the highest fitness. Note that a failure to observe these changes would not have been taken as evidence against natural selection in nature. In fact, no changes were observed in cages kept at 16°C, an observation that was folded into the explanatory scheme as an example of how fitnesses of genotypes are sensitive to environment.

The second route of demonstration of natural selection was to show that, *in nature*, there was a consistent excess of heterozygous adults above the expected Hardy-Weinberg proportions, a demonstration that, in the words of Dobzhansky and Levene (1948), was "Proof of the operation of natural selection in wild populations of *Drosophila pseudoobscura*." Again, the demonstration depends in no way on any ecological or physiological observations but entirely on the deviation of observed from theoretical frequencies. And again, the failure to find a significant deviation from Hardy-Weinberg frequencies could not have been taken as disproof of the operation of natural selection since a variety of selective mechanisms including frequency-dependent selection (a favorite of Sewall Wright) and fertility differentials would not necessarily be detectable in this way (see Prout 1965). Both the laboratory cages and the genotypic frequencies from nature illustrated what Dobzhansky and Wright had already concluded from their observations of frequency patterns in nature.

Over and over again, Dobzhansky performed experiments to exemplify general principles without attempting to cash them out in detail for specific cases. A number of experiments showed how fitness of genotypes depended on environment (Dobzhansky and Spassky 1944, GNP XI; Dobzhansky et al. 1955, GNP 23; Dobzhansky and Levene, 1955, GNP 24), but the environments chosen for these demonstrations, three different temperatures and three different species of yeast food not natural to the flies, were not meant to explain in material detail the fate of genes in nature; they were simply different environments chosen to illustrate the general principle that each genotype had its unique norm of reaction, that one could not predict how well a genotype would be in the environment from its performance in another environment, and that heterozygotes were generally less susceptible to environmental fluctuations than homozygotes. There was never any question whether these propositions were true in general. The empirical problem was to illustrate them in practice. The "experiments" were demonstrations.

The other chief preoccupation of Dobzhansky's empirical work was the estimation of parameters such as mutation rates, migration rates, the size of the "panmictic unit" or, more generally, the hierarchical structure of F-statistics, in order to fit the observations of natural genetic variation into the analytic apparatus created by Sewall Wright in his general theory of evolution in Mendelian populations (Wright 1931). It is important to understand that Wright's "theory" is untestable in the classic sense, because it is not a "theory that" but a "theory of," that is, a synthesis of all the relevant forces operating on gene frequencies in populations. The parameters of that theory, especially the parameters of breeding structure, are not estimated autonomously and then put into the structure. Rather, the observations of frequency patterns are themselves the data for a tautological estimation of the parameters, given the theoretical apparatus. Indeed, we do not even know what we mean by "effective breeding size" in a natural population except as a dual representation of the actual degree of coancestry among individuals. Many of Dobzhansky's measurements of these parameters cannot be understood except as the filling-in of quantitative values in an unquestioned theoretical apparatus.

Dobzhansky's relations with theoreticians were rather like those of an architect building a skyscraper to structural engineers. He understood intuitively what the theory entailed but depended upon the practitioners to provide the technical expertise to turn the intuitive theory into a structure that would stand up. Dobzhansky depended upon them absolutely (and therefore, perhaps, resented them, too) for the detailed planning and analysis of the experiments. There is, for example, a plaintive letter from Dobzhansky to Wright, dated May 14, 1941, in which he begs Wright to tell him what the correct experimental plan is to be for the summer work at Mather in which dispersal of *D. pseudoobscura* is to be measured. Moreover, to Wright he owed his original understanding of population structure and the interaction between drift and selection. (The suggestion has been made that, in fact, A. H. Sturtevant first showed Dobzhansky the importance of Wright's theory for work on natural populations. See Provine's essay in Lewontin et al. 1981 and letters from Dobzhansky to Wright in the early 1930s from Pasadena in the Wright-Dobzhansky correspondence. Whether or not Sturtevant's role was crucial, there is no doubt that Dobzhansky developed an extraordinarily subtle understanding of Wright's theory, despite his own lack of mathematical facility.)

However, Dobzhansky never depended upon theoreticians for the conceptualization of the experiments and never allowed them to influence his interpretation of the results. This last point is material to the claims that Dobzhansky was a "theoretician without tools." Dobzhansky knew at all times what the experiments were intended to demonstrate and what the conclusion was from the results. Indeed, if my claim is correct that the experiments were largely *demonstrations* rather than explorations, these conclusions were already in existence *before* the experiments were done. The role of the theoretician was to use his expertise to draw the rigorous quantitative connection between the data and the conclusions that Dobzhansky had already decided upon. Any real independence on the part of the theoretician to draw other conclusions from the data was not tolerated, and this applied even to Sewall Wright, whom Dobzhansky believed to be "one of the two greatest living geneticists." (The other was Muller.)

Two incidents illustrate the point. In 1951-52, A. R. Cordeiro, working at Columbia, carried out a series of experiments, planned by Dobzhansky, on chromosomal heterozygotes and homozygotes in *D. willistoni*, designed to demonstrate the fitness superiority of heterozygotes (Cordeiro 1952). Howard Levene was asked to carry out the statistical analysis of a fairly complex experimental design, and his first analysis showed the expected superiority of heterozygotes and their lower variation in fitness. Shortly thereafter, Levene reworked the analysis and reported to Dobzhansky that his second, better analysis failed to show the results expected. Dobzhansky was exceedingly annoyed and said the paper could not be published. He pressured Levene into yet another analysis, which now showed significant results. These were immediately incorporated into the paper, and without further ado, the paper was published.

Yet more extraordinary were the circumstances of publication of GNP 07 (Pavlovsky and Dobzhansky 1966). The population cage experiments were analyzed by W Anderson, who was originally one of the authors of the paper, and the conclusion was the conventional one of fitness superiority of inversion heterozygotes. Wright was a reviewer of the paper and carried out his own analysis, concluding that there was actually frequency-dependent fitness rather than unconditional superiority of heterozygotes. On reading Wright's review, Anderson concurred, but Dobzhansky refused to change his conclusions. Anderson was dropped from authorship, although the results of his original calculations were used in the paper, and Dobzhansky simply reiterated his already predetermined conclusion. (See discussion of this incident in Lewontin et al. 1981:807-8 for details.)

It should not be concluded that Dobzhansky was incapable of rejecting a long-held theory in the face of overwhelming evidence. On the contrary, he twice abandoned strongly held theoretical commitments. The first, in 1941, was the change from his original belief that the inversions of *D. pseudoobscura* were entirely subject to random genetic drift to his unshakable conviction after that time that the inversion frequencies were determined by natural selection (See Lewontin et al. 1981:303). The second was his abandonment in 1953 of the theory of coadaptation, in favor of the belief in the unconditional superiority of allelic heterozygotes. I now turn my attention to this later change.

By the time of the appearance of the third edition of *Genetics and the Origin of Species* in 1951, Dobzhansky had formed a well-articulated theoretical view. This view included the following elements:

1. The fitnesses of genotypes, like other aspects of their phenotype, were contingent on environment and were generally unpredictable from one environment to another.

2. There is no "wild-type" genotype because all populations of sexually reproducing organisms are characterized by a large amount of genetic variation, so that the "typical" individual is necessarily a heterozygote at many loci.

3. Homozygotes *from natural populations* were less fit than heterozygotes from natural populations in most environments, but there are environments in which particular homozygotes may be superior. Homozygotes were "narrow specialists" while heterozygotes were "broad generalists" in fitness.

4. In particular, inversion heterozygotes had higher fitness than inversion homozygotes *in the natural populations from which those chromosomes were taken.*

5. The superiority of heterozygotes over homozygotes is itself a consequence of natural selection and is not some intrinsic property of heterozygosity per se. That is, populations will accumulate, merely as a mechanical consequence of the stable equilibrium of gene frequencies caused by fitness overdominance, a set of alleles at a locus that show overdominance in fitness when alleles will be either fixed or lost by natural selection and drift. This is the theory of *coadaptation.*

A long-known result of simple population genetic theory was that if the fitnesses of homozygotes AA and aa were only $(1-s)$ and $(1-t)$ while the heterozygote had relative fitness 1.0, then a stable equilibrium of allele frequency was produced with $p(A) = t/(s+t)$. If, on the other hand, the homozygotes were more fit than heterozygotes, there would be fixation of allele frequencies at 1 or 0. Dobzhansky reasoned that every population was constantly undergoing mutations and that those mutant alleles that had the heterotic property would accumulate in the population, But since heterosis is a *relational* property between two alleles, then which alleles were accumulated would depend on which other alleles were present at a particular time in the life of the population.

This historical contingency in conjunction with the random events associated with finite population size and finite population lifetime would mean that the particular ensemble of heterotically maintained alleles in a population would be different from the ensemble in a different population. There was, therefore, no reason to suppose that a random allele from population A would be heterotic in combination with a random allele from population B. This contingency would be even greater because of epistatic effects between loci. Since fitness of a genotype depended on environment, it also depended on the genetic environment at other loci. Whether or not a heterozygote $A_i A_j$ was heterotic would depend, not only on the identity of $A_i A_j$ but also on the alleles at other loci that were historically present. Thus Dobzhansky pictured a process of selective coadaptation taking place during the history of a population, the coadaptation of alleles at a locus to each other, and of the alleles at one locus to the alleles at other loci. These two elements of *coadaptation* are parallel to Mather's notions of *relational balance* and *internal balance* in artificially selected populations. The alleles at loci along a haploid chromosome would be internally balanced to form a superior linked unit, and this unit would, in turn, be relationally balanced with other haplotypes in the population.

Dobzhansky's theory of coadaptation included the inversion heterozygotes and homozygotes. The fitness properties of inversion types, according to Dobzhansky, were the result, not of the inversion breaks themselves, but of the allelic content of the inverted chromosomes locked up in nonrecombining blocks. It followed, then, that an inversion heterozygote, say, ST/CH, would be superior in fitness to the homozygotes ST/ST and CH/CH only if the ST and CH chromosomes were taken from the same natural population where the allelic contents of the two gene pools had become coadapted by selection. If ST and CH were taken from two different populations, the alleles would have no selective history relative to each other and there would be no heterosis. GNP 19 (1950) was to be a demonstration of this truth. As the title "Origin of heterosis through natural selection in populations of *Drosophila pseudoobscura*," suggests, the result was as expected.[1] Whereas several previous experiments (Wright and Dobzhansky 1946; Dobzhansky 1947) had shown a superior fitness of inversion heterozygotes, when inversions from Mather, California; Piñon Flats, California; and Chihuahua, Mexico, were crossed with one another, the heterozygotes were always either intermediate or even inferior in fitness to chromosomal homozygotes. This observation held without any exception for all combinations of different inversions, ST, AR, and CH, both for larval viability and in net fitnesses in population cage experiments. ST chromosomes from Piñon Flats had no selective experience of AR chromosomes from Mather, so no coadaptation had occurred. This experiment is a classic in the "demonstration" mode and was a particular triumph for Dobzhansky who had wagered and won a dime, framed and displayed in his office, against the contrary prediction of H. J. Muller.

The next step in the demonstration of the universality of coadaptation was to carry it to chromosomes not involved in inversion polymorphism. By 1951, it was universally agreed that if chromosomes are sampled from a natural population and made homozygous, their carriers would generally be of lower viability and fertility than random reconstituted heterozygotes from those same chromosomes. No experiment had ever given a different result. (See, for example, GNP 8, Dobzhansky, Holz, and Spassky 1942; GNP 11, Dobzhansky and Spassky 1944; and later experiments such as GNP 22, Dobzhansky and Spassky 1954.) The theory of coadaptation predicted, however, that if heterozygotes are made between chromosomes from geographically distant populations, no heterosis would be seen. No such experiment had been done, but Dobzhansky was certain of its outcome and assigned the demonstration to M. Vetukhiv, an exiled Ukrainian geneticist who had a fellowship to work at Columbia.

---

[1]   After completing the present paper, I became acquainted with Jane Maienschein's very fruitful concept of "epistemic styles" (Maienschein 1981). What has been illustrated here is an epistemic style that was of great influence on the future of population genetics.

The experimental design was meant to eliminate the effect of homozygosity for rare recessive deleterious genes that is always the result of mating isogenic chromosomal homozygotes. That is, the previous experiments, cited above, always contrasted totally homozygous chromosomes with random heterozygotes. But a total homozygote would, on any theory, show inbreeding depression because it is almost certain that any chromosome in a natural population carries a few rare recessive deleterious genes. The Vetukhiv experiment, then, did not make chromosomal homozygotes but contrasted the fitness components of crosses between chromosomes taken from the same geographical population (*intrapopulation* crosses) with crosses between chromosomes from different populations (*interpopulation* crosses). These crosses were then carried to the F_2 to show the effects of recombination of genes from different populations, breaking up the internally balanced haplotype. Several sets of experiments were carried out, measuring viability (Vetukhiv 1953, 1954), one measuring fecundity of females (Vetukhiv 1956), and one measuring longevity (Vetukhiv 1957). A fourth set of a more complex experimental design was carried out by D. Brncic (1954).

The results of the Vetukhiv experiments produced a major intellectual crisis for Dobzhansky. Instead of the interpopulational hybrids being intermediate or lower in fitness than intrapopulational crosses, heterosis was observed in all components of fitness for all five of the populations involved. There did appear to be fitness breakdown in the F2 recombinant generation, as predicted from coadaptational theory, but the clear heterosis in F1s between chromosomes that had never experienced joint selective histories in nature was a fatal blow to coadaptation. There was some skepticism in the laboratory at Columbia about the experimental design because the parental (intrapopulation), F1, (interpopulation), and F1 crosses were not made and measured simultaneously, This was partly remedied by carrying out a new set of experiments in which parental and F1 crosses were simultaneous and a separate set of parental and F2 crosses were made simultaneously. The results were the same. Coadaptation was dead and in its place was the extraordinary result that any heterozygote was likely to be superior in fitness to its constituent homozygotes, irrespective of joint selective history and eliminating the effect of rare deleterious mutants.

Dobzhansky's reaction was to encourage more experiments by Vetukhiv and Brncic and to convert his own theoretical view from coadaptation to a radical belief in the fitness superiority of heterozygotes per se. From 1955 onward, only five years after the triumphant outcome of the wager with Muller, the theory of coadaptation disappears from the central dogma of the Dobzhansky school. In its place are an equally assured and unquestioned adherence to a belief in the intrinsic superiority of heterozygotes and a long sequence of demonstration experiments in support of this view (for example: Dobzhansky and Levene 1955, GNP 24; Spassky et al. 1960, GNP 29; Dobzhansky, Krimbas, and Krimbas 1960, GNP 30; Wallace and Dobzhansky 1962, Dobzhansky, Spassky, and Tidwell 1963, GNP 32). (Typical of a demonstration experiment, the paper by Wallace and Dobzhansky 1962 is entitled "Experimental *proof* of balanced genetic loads in *Drosophila*" (emphasis added). It is reminiscent of GNP 17, "Proof of the operation of natural selection in wild populations of *D. pseudoobscura.*" A rhetorical analysis of Dobzhansky's paper titles is beyond this essay but would be well worthwhile.) The subsequent history of the struggle between the "balanced" and "classical" schools over the importance of heterosis and overdominance is also not in the scope of the present paper (but see Lewontin 1974 for one version).

What is the explanation of Dobzhansky's conversion from one theoretical dogma to another? Certainly the Vetukhiv experiments were a potent force. We might say that while Dobzhansky was ready to discount *alternative* explanations of phenomena that did not agree with his theoretical prior position, even when those who disagreed with him were of the highest intellectual level in his pantheon (S. Wright and H. Muller, for example), he was not prepared to sweep observed *contradictions* under the rug. But the distribution between alternative and contradictory is not so clear. All scientists must be prepared to explain away apparently contradictory observations in the face of well-grounded and strongly held belief. Dobzhansky was not an exception, and he explained away contradictory results from other laboratories as experimental artifacts. Given my claim that Dobzhansky was essentially a theoretical scientist, engaged, not in experimental tests, but in experimental demonstration, it is more likely that Dobzhansky was already in the process of a *theoretical* shift to which the Vetukhiv experiment gave a final push.

That theoretical shift was a result of the influence of M. Lerner and his developing ideas of "genetic homeostasis." Lerner had a powerful personal influence on Dobzhansky who felt very close to him. When *Genetic Homeostasis* (1954) appeared, Dobzhansky took it up immediately and it became central to laboratory discussion and to the training of graduate students. Lerner's view was that heterozygotes have the property of greater developmental and physiological buffering against disturbance of the adapted phenotype and that this greater buffering, an *intrinsic* property of heterozygosis, lay at the base of the higher fitness of heterozygotes. Such a theory was consonant with and unified all the elements of Dobzhansky's theoretical apparatus, except for the theory of coadaptation (see above), and at the same time predicted Vetukhiv's observations. Its theoretical power must have been irresistible to Dobzhansky. Especially, it provided a general explanation of the central core of Dobzhansky's observation of nature, the genetic heterogeneity among individuals that Dobzhansky regarded as the critical fact of biology.

To make the theory of intrinsic heterozygous superiority the central theoretical commitment of his view, Dobzhansky had to give up a general theory of the operation of natural selection, one for which he had offered a final triumphant demonstration only five years before. It also involved putting out of consideration the observations on which the theory of coadaptation was based and which were, in turn, contradictory to the new theory. So Dobzhansky's science changed, not by the progressive reduction of contradictions between observations and theory, but by the choice of a theoretical apparatus on the basis of quite other considerations that are yet to be fully understood.

References: The references for this paper by R. C. Lewontin will be found in the book by L. Levine.

DROSOPHILA PSEUDOOBSCURA[1]

REVIEWED BY A. H. STURTEVANT

Reprinted with the permission of Ecology

This volume contains three papers. The first, by Dobzhansky and Epling, deals with the three American representatives of the group-Drosophila miranda, D. pseudoobscura Race A, and D. pseudoobscura Race B. Their geographical distribution is outlined, and a general account of their ecology is given. There follows a discussion of detailed observations made on Race A in the region of Mt. San Jacinto, in southern California. A series of ingenious and laborious field experiments was carried out, over two successive years, in studying length of life, rates and distances of movement of adult flies, and the effects of light, temperature, and humidity on these activities. The authors recognize that some of the determinations are only approximate, and that most of the values obtained are applicable only to the season and locality studied; nevertheless we are here presented with a detailed picture of the life of this form under natural conditions that is far more precise than anything hitherto available for any Drosophila.

---

[1] Dobzhansky, Th., and Carl Epling. 1944. Contributions to the genetics, taxonomy, and ecology of *Drosophila pseudoobscura* and its relatives Carnegie Institution of Washington, Publication 554. Pp. iii + 183.

Dobzhansky and Epling describe a new species, *Drosophila persimilis*, using the name for the form long known as *D. pseudoobscura* Race B, the name *pseudoobscura* being restricted to the form generally known as Race A. The reviewer is inclined to regard *persimilis* as a synonym of *D. lancefieldi* Ginsburg—a conclusion avoided by Dobzhansky and Epling by what seems to me to be a clear misinterpretation of Ginsburg's intention, and by failure to recognize that he does give "an unequivocal bibliographical reference." The necessity for the use of a new specific name at all is at least open to question; one of the reasons given is that "Reeds' wing index," used in conjunction with other characters, probably makes it possible to distinguish the two forms on purely morphological grounds. This index is obtained by multiplying the wing area by the cube of wing length (Reed, Williams and Chadwick, Genetics **27**: 349-361; 1942). It was shown by Mather and Dobzhansky (Amer. Nat. **73**: 5-25; 1939) that wing length and wing width are highly correlated; this index, therefore, is nearly equivalent to the fifth power of wing length—it is, in short, a method of greatly magnifying slight differences in wing length, with no increase in the significance of such differences. It was shown by Mather and Dobzhansky (loc. cit.) that there is no consistent difference between the two races in question in wing length—the averages for the two sexes even differ in opposite directions. The index is, therefore, useless in distinguishing Race A from Race B. The name *persimilis* being of doubtful validity, and being based in part on a nonexistent character difference, I shall use the older designations of Race A and Race B in this review. One advantage of this system is that it gives a name (*pseudoobscura*) to the complex made up of Race A and Race B—which is as far as one can go in identifying wild specimens without breeding from them or examining their chromosomes. Another advantage is that this is the established system, in use in the current literature; and the paper under review does not replace that literature, since it is not a summary of existing information on the group.

The second paper, by Dobzhansky, gives the data on the nature and occurrence of the gene sequences in the two races of D. pseudoobscura. Most of the discussion centers on the sequences found in chromosome III—of which Race A has 15, and Race B, 7. One sequence ("Standard") is common to the two, so that a total of 21 sequences is known. It is to be observed that the present account, since it does not include the detailed figure of the salivary gland chromosome bands, will not enable the student to recognize the various sequences in his own preparations for that purpose reference must be made to earlier literature. This work has been carried out on a tremendous scale—the author estimates that he has made cytological determinations on the sequences present in 20,000 third chromosomes. There results a picture of the local and seasonal diversities within populations, and of the geographical ranges and local frequencies of individual sequences, that is far more detailed than that available for any other organism. This material is utilized for an analysis of the factors involved in population questions as the degree of inbreeding, rate of migration between populations, the effects of such ecological barriers, the distances required to make possible the development of genetic diversities, and the causes of seasonal changes in relative frequencies of sequences at one locality. In connection with the last-named problem, the reviewer finds the conclusions reached by Dobzhansky difficult to accept. A population living at the edge of the Colorado Desert is found to show a regular seasonal fluctuation, such that the "Chiricahua" sequence is present in about 15 per cent of the chromosomes from September through March, but then increases to about 35 per cent in June. (See also fuller discussion by Dobzhansky in Genetics 28:162-186; 1943.) Dobzhansky interprets this as an effect of balanced selective agents—the Chiricahua sequence is selected against during the desert summer, but is not eliminated because selection favors it during April and May. Opposed selective actions of the magnitude required seem very unlikely; a more probable interpretation would seem to be that the lower frequency is characteristic of the permanent breeding desert population, while the higher one is due to the presence of migrants from the adjacent mountain—which has very large populations at this season, with about 40 per cent Chiricahua chromosomes. These migrants are, however, unable to withstand the desert summers, and, therefore, by September the frequency of Chiricahua has again returned to 15 per cent. One who has experienced the down-hill winds in this region will recognize a possible reason for the postulated migration. The suggested interpretation is in agreement with the fact that the Arrowhead sequence constitutes about 25 per cent of both populations at all times. Migration should, therefore, bring about no change in its frequency, whereas on Dobzhansky's interpretation its immunity to selective effects adds a serious complication.

The 21 known sequences can be arranged in a phylogenetic tree that is far more unambiguous than the usual hypothetical pedigree. The third paper, by Epling, is a discussion of the relation between this phylogeny and the known geographical distribution of the sequences. The conclusion is reached that many of the present sequences existed in Miocene time, and that the present distribution is to be explained largely in terms of climatic conditions at that time and since. To the reviewer it seems that at least two of the primary bases of the argument are very questionable. Much is made of the discontinuous distribution of the "Santa Cruz" sequence (from Oregon to Lower California, and from Chihuahua to Guatemala). This discontinuity seems to the reviewer to be not established. Santa Cruz occurs along the western edge of the Colorado Desert, and has been found in the northern part of the desert itself. It also occurs on Cedros Island, off the middle of Lower California. It may well be found in the area between these localities and central Chihuahua—an area in much of which the species is certainly rare, although it is probably nearly everywhere present, and from which very few data are available.

Epling recognizes that the gene arrangements themselves probably have no selective value, yet he supposes that climatic factors may exercise selective effects on them. The mechanism of such selection is not apparent. There is evidence that can be interpreted as meaning that in local populations there may be minor differences between the sequences, that lead to selective effects, though, as indicated above, the reviewer favors a different interpretation of the data. Even if such differences exist, they are surely of a temporary and local nature. In the San Jacinto region, the Standard sequence is apparently favored by dryer and warmer times of year, at the expense of Chiricahua; yet, taking into account the general geographical distribution of the two sequences, the reverse relation is what would be expected—and the data suggest that the relation is in fact reversed in the San Gabriel canyon, about 70 miles from San Jacinto.

There is a real historical problem involved in the geographical distribution of the sequences; but perhaps not enough data are yet available for its solution. Certainly there are several regions from which more information might be very illuminating—such regions as northwestern and west central Mexico, northern Utah, and the adjacent portions of Nevada, Idaho, and Wyoming.

A. H. Sturtevant
California Institute of Technology, Pasadena

MENDEL, LINKAGE, AND SYNTENY

Reprinted from BioScience

E. Novitski, University of Oregon, and Styg Blixt, Plant Breeding Institution, Landskrona, Sweden

It is a common misconception, particularly among non-pea geneticists and teachers of elementary genetics, that since Mendel did not encounter any deviations from independent assortment, the seven pairs of alleles that he worked with must have been distributed evenly. Thus there would be only one pair on each of the seven chromosomes of Pisum sativum, making him lucky indeed not to have worked with eight pairs for then linkage would have been inevitable. The complications of linkage would have so disturbed Mendel—so goes the conventional wisdom—that he would have doubted the generality of independent assortment and would probably never have published his work.

This argument is not correct. The likely locations of the factors Mendel used have been discussed by a number of workers (Lamprecht 1968a, b, Nilsson 1951) but, unfortunately, have been published in non-English journals or books. Not only is it not generally appreciated that this matter has been discussed in detail, but it is also not widely known that there exists some disagreement among the authorities in the field of pea genetics with respect to the exact allocations.

To counter this misunderstanding more effectively, one of us (Blixt 1975a) has published a short note in the English language pointing out that, in fact, both chromosomes I and 4 of the garden pea each carry more than one of Mendel's seven pairs. The map distances separating them, however, are so great that he could not have detected linkage without using intermediate loci. In fact, two loci that he tested extensively together, I-i and A-a (or B-b and C-c, respectively, by his symbolism) are on chromosome 1, but the great genetic distance separating them (204 units) precluded his detection of linkage.

There is room for debate on the matter of the assignment of Mendel's locus for smooth vs. wrinkled pods. In the published works just cited and among plant geneticists, there are two opposing points of view. One of these, espoused by Lamprecht (1968a, b) and Blixt (1975a, b) is that this locus is on chromosome 4. The second, favored by Nilsson (1951) as well as others,[1] is that it is on chromosome 6. In this note, we shall present some of the arguments for each of these points of view.

ARGUMENTS FOR CHROMOSOME 4

If the characters Mendel used with this phenotype correspond to the V-v locus of the pea (at 211 on chromosome 4), then it would have been only 12 units from the plant height locus, Le-le (at 199 on chromosome 4); Mendel, therefore, might have encountered linkage. Although the specific cross involving plant height and pod form was not one that Mendel reported quantitatively, he did state at two different points in his paper (Mendel 1866) that he made all possible (128) combinations of the seven pairs of characters he worked with; this would have involved getting crossover types not just once or twice, but repeatedly.

Furthermore, in his letter to Nägeli on 18 April 1867 (Stern and Sherwood 1966), he mentions explicitly that during the period from 1859 he raised many plants from the hybrid between two lines that differed (among other characteristics) by plant height and pod form. The motivation for continuing this line seems to have been more gustatory than scientific because the hybrid had large tasty seeds ("grossen wohlschmeckenden Samen"). Close linkage might have made it difficult for him to obtain a nonsegregating DDGG combination from the initial cross of ddGG by DDgg. We would like to imagine that the persistent production of the recessive phenotypes during continuous inbreeding for the desirable dominant characteristics would not have escaped his attention.

The crossover values between the Le and V loci are quite variable; although they ordinarily average about 10%, they can vary from 2.6 to 38.5% (Lamprecht 1968a). If his strains happened to have crossover values on the higher side or a high mutation rate,[2] then of course the problem of linkage would diminish or even disappear.

ARGUMENTS FOR CHROMOSOME 6

There is, however, the second possibility, which is that the characters used for the form of the pods is not the V-v locus on chromosome 4, but the P-p locus on chromosome 6, as suggested by Nilsson as early as 1951 and again in 1967. However, only one variety homozygous for p seems to have existed in Mendel's day, Sugarpea de Grace (Buchsbaum). This was genetically le, V, p, a weak variety, which was recommended only for greenhouse cultivation. Unfortunately, Mendel never seems to have named the varieties he used, and this information does not appear to be available from the records at Brno.[3]

This might be determined, given enough time and effort, by a search through the records of H. Lamprecht, who grew all of the varieties Mendel is supposed to have used, or by a study of the 19th century German seed catalogues. However, the great extinction of old varieties in Europe during the 1940s and 1950s makes it almost certain that the original strains are not now in existence. We, therefore, doubt that absolute certainty can ever be reached on this point; both alternative locations for the pod character D-d of Mendel are possible.

PROBABILITIES OF VARIOUS DISTRIBUTIONS

On the assumption that all seven pea chromosomes are approximately equal in length, the chance that a random sample of seven loci will be distributed two on each of two chromosomes and one on each of three others (the configuration with Mendel's D corresponding to P) is 0.32, the most likely of all possible distributions. A similar calculation for the distribution of three on one chromosome, two on another, and the remaining two occurring singly (representing the case where Mendel's D is in fact V) gives a probability of 0.21, the second most likely possibility, On the other hand, the chance that all seven would fall on different chromosomes is 0.006. A more elegant calculation, which takes into account the fact that two of the chromosomes are somewhat shorter than the others, decreases this figure to about 0.004.[4] From this point of view, it would seem that those who were (mistakenly) astonished at Mendel's unusual luck in not extending his work to involve an eighth locus, making linkage inevitable, should, instead, have been concerned with the improbability of randomly selecting seven loci that happened to fall each on a different chromosome!

LINKAGE vs. SYNTENY

In conclusion, this discussion indicates the usefulness of the term "synteny" proposed by Renwick (1971) to designate the occurrence of two loci on the same chromosome, independent of distance. *Linkage* is now used ambiguously, first, implying genetic evidence of a departure from independent assortment but then also used to designate location on the same chromosome. *Linkage* might well be limited to the first meaning, and *synteny* used for the second. With chromosomes as genetically long as those of the garden pea (average length of about 180 units), the chance of detecting linkage, even with synteny, is rather small. If the common misstatement about Mendel's not encountering linkage is expressed instead in terms of his failure to establish synteny, the weakness of the argument becomes self-evident.

[1] Personal Communication from A. Gustafsson, Institute of Genetics, University of Lund, Sweden, and R. Lamm, Institutionen fur Trädgärdsvetenskap och Landskapsplanering, Alnarp, Sweden.

[2] Although one would not ordinarily consider reverse mutation a possible source for pseudorecombinants, occurring with a high frequency, for this locus unusually high mutation rates have been reported. Average values are reported to run at about 0.2 to 1%, but in some cases run as high as 40% (See Lamprecht 1941)!

[3] V. Orel, Moravske Museum, Mendelianum, Brno, Czechoslovakia, personal communication.

[4] We are indebted to Lee Douglas, Nijmegen University, The Netherlands, for this more exact calculation.

REFERENCES

Bauer, H. and Th. Dobzhansky. 1936. A comparison of gene arrangements in Drosophila azteca and D. athabasca. (Abstr.) Rec. Genet. Soc. Amer. 5; and 1937. Genetics 22:185.

Blixt, S. 1975a. Why didn't Gregor Mendel find linkage? Nature 256: 206.

————. 1975b. The pea. Pages 181-221 in R. C. King, ed. Handbook of Genetics, vol. 2. Plenum Press, New York

Carlson, E. A. 1966/1989. The gene: a critical history. W. B. Saunders/Iowa State University Press, Philadelphia and London/Ames, Iowa.

Carlson, E. A. 1971. An unacknowledged founding of molecular biology: H. J. Muller's contribution to gene theory. 1910-1936. Journal of the History of Biology 4(1):149-170.

Carlson, E. A. 1974. The Drosophila group: the transition from the Mendelian unit to the individual gene. Journal of the History of Biology 7:31-48.

Cooper, K. W. 1959. Cytogenetic analysis of major heterochromatic elements (especially Xh and Y) in D. m. and the theory of "heterochromatin." Chromosoma 10:535-588.

Correns, C. 1900. Mendel's Regel über das Verhalten der Nachkommenschaft der Rassenbastarde. Berichte der deutschen botanischen Gesellschaft 18:158-168. English translation by Piternick reprinted in Stern and Sherwood (1966).

Crow, J. F. 1987. Muller, Dobzhansky, and overdominance. Journal of the History of Biology 20: 351-380.

Crow, J. F. 1993. Felix Bernstein and the first human marker locus. Genetics 133:4-7.

Darwin, C. 1868. in The works of Charles Darwin: The variation of animals and plants under domestication, Vol. II Charles Darwin, Paul H. Barrett (Editor), R. B. Freeman (Editor), Peter Cautrey (Editor) New York University Press; New York, (December 1989).

Dobzhansky, Th. 1935. Drosophila miranda, a new species. Genetics 20:377-391.

Dobzhansky Th. 1937. Genetics and the origin of species. 1st edition. Columbia University Press New York.

Dobzhansky Th. 1962. Oral history memoir. Columbia University Press. Available in microfiche only.

Dobzhansky, Th. and C. Epling. 1944. Taxonomy, geographic distribution, and ecology of D. pseudoobscura and its relatives. Publ. Carneg. Instn. [Wash.], No. 554:1-46.

Dobzhansky, Th. and O. Pavlovsky. 1953. Indeterminate outcome of certain experiments on Drosophila populations. Evolution 7:198-210.

Dobzhansky, Th. and M. L. Queal. 1938. Genetics of natural populations. I. Chromosome variation in populations of Drosophila pseudoobscura inhabiting isolated mountain ranges. Genetics 23:463-484.

Dobzhansky, Th. and D. Socolov. 1936. Structure and variation of the chromosomes in D. azteca. J. Hered. 30:3-19.

Dobzhansky, Th. and C. C. Tan. 1936. Studies on hybrid Sterility III. A comparison of the gene arrangement in two species, Drosophila pseudoobscura and Drosophila miranda. Z. indukt.Abstamm. u. VererbLehre. 72:88-114.

Dutch Book see Morgan, Bridges and Sturtevant 1925.

Fabergé, A. C. 1986. Documenting eugenics and human genetics. Journal of Heredity 77:287-288.

Edward, A. W. F. 1986 Are Mendel's Results Really Too Close? Biol. Rev. 61:295-312.

Feynman R. P. 1999. The pleasure of finding things out. Perseus Publishing, Cambridge, Mass.

Fisher, R. A. 1936 Has Mendel's Work Been Rediscovered? Annals of Science 1:115-137

Gardner, H. 1999. Intelligence reframed. Basic Books, New York City.

Goodstein, J. R. 1991. Millikan's school, a history of the California Institute of Technology. W. W. Norton Co. N. Y.

Hardy, G. H. 1940. A mathematician's apology. Cambridge University Press.

Hartl, D., Y. Hiraizumi and J. F. Crow. 1967. Evidence for sperm dysfunction as the mechanism for segregation distortion in D. m. Proc. Nat. Acad. Sci., Wash. 58:2240-2245.

Lamprecht, H. 1941. Über Gentabilität bei Pisum. Zuchter 13: 97-105.

Lamprecht, H. 1968a. Die neue Genenkarte von Pisum und Warum Mendel in seinen Erbsen-Kreuzungen keine Genekoppeling gefunden hat. Arb. Steinermärkischen Landesbibliothek Joanneum, Graz.

Lamprecht, H. 1968b. Die Grundlagen der Mendelschen Gesetze. Verlag Paul Parey, Berlin-Hamburg.

Lancefield, D. E. 1925. An interracial cross of Drosophila obscura producing partially fertile hybrids. (Abstr.) Anat. Rec. 13:346.

Lancefield, D. E. 1929. A genetic study of crosses of two races or physiological species of Drosophila obscura. Z. indukt. Abstamm.-u. VererbLehre 52:287-317.

Levine, L. 1995. Genetics of natural populations. Columbia University Press, New York.

Lewis, E. B. 1939. Star-recessive, a spontaneous mutation in D. m. Proc. Minn. Acad. Sci. 7:23-26.

Lewis, E. B. 1995. Remembering Sturtevant. Genetics 141: 1227-1231.

Lewontin, R. C. 1995. Theodosius Dobzhansky—a theoretician without tools, see Levine (1995), also as appendix in this book.

L'Héritier, P. and G. Teissier 1934 Une expérience de sélection naturelle. C. R. Soc. Biol. Paris 117:1049-1051.

Lindsley, D. L. and E. Novitski. 1950. The synthesis of an attached X-Y chromosome. Dros. Inf. Serv. 24:84-85.

Lucchesi, J. C. 1994. Sturtevant's mantle and the (lost?) art of chromosome mechanics. Genetics 136:707-708.

MacKnight, R. H. 1939. The sex-determining mechanism of D. miranda. Genetics 24:180-201.

Mendel, G. 1866. Versuche über Pflanzen-Hybriden. Verhandlungen des naturforschenden Vereines in Brünn 4:3-47. English translation first published in 1901 in Journal of the Royal Horticultural Society 26:1-32. Also an English translation by Sherwood in Stern and Sherwood 1966.

Metz, C. W. and M. S. Moses. 1923. Chromosomes of Drosophila. A comparison of the chromosomes of different species of Drosophila. J. Hered., 14:195-204.

Miller. D. D. 1939. Structure and variation of the chromosomes in D. algonquin. Genetics 24:699-708.

Moran, P. A. P., E. Novitski and C. E. Novitski. 1969. Paternal age and the secondary sex ratio in humans. Ann. Hum. Genet. 32:315-316.

Morgan, T. H., C. B. Bridges and A. H. Sturtevant. 1925. The Genetics of Drosophila, Bibliographica Genetica II: 1-362.

Nilsson, E. 1951, Trädsgardsärter. Svensk Vaxförädling, Stockholm, Sweden.

Nilsson, E. 1967. Ärftlighetslärans urkunder. Corona, London.

Noor, M. A. F., N. A. Johnson, and J. Hey 2000. Gene flow between Drosophila pseudoobscura and D. persimilis. Evolution 54,6:2174-2175.

Novitski, C. E. 1995. A Closer Look at Some of Mendel's Results. Journal of Heredity 86(1): 62-66.

Novitski, E. 1938. Heldout, a recessive wing mutation in Drosophila melanogaster. Proc. Ind. Acad. Sci. 47: 256-260.

Novitski, E. 1946. Chromosome variation in D. athabasca. Genetics 31:508-524.

Novitski, E. 1949. An inversion of the entire X-chromosome. Dros. Inf. Serv.23:94-95.

Novitski, E. 1950. The transfer of mutant genes from small inversions. Genetics 35:249-252.

Novitski, E. 1951. Nonrandom disjunction in Drosophila. Genetics 36:267-280.

Novitski, E. 1953. The dependence of the secondary sex ratio in humans on the age of the father. Science 117:531-533.

Novitski, E. 1954. Sex ratio and parental age. Science:119:473-474.

Novitski E. 1976. ABO blood groups and the Hardy-Weinberg equilibrium. Science 191:478.

Novitski E. 1983. Genes, radiation and society. American Scientist:71:527-528.

Novitski, E. and E. R. Dempster. 1958. An analysis of data from laboratory populations of Drosophila melanogaster. Genetics 43:470-479.

Novitski, E. and A. W. Kimball. 1958. Birth order, parental ages, and sex of offspring. Amer. Jour. of Hum. Genetics:10:268-275.

Novitski, E. and D. L. Lindsley. 1950. Construction of tandemly attached-X chromosomes. Dros. Inf. Serv. 24:90.

Novitski E. and S. A. Rifenburgh. 1938. Heldout, a recessive wing mutation in Drosophila melanogaster. Proc. Ind. Acad. Sci., 47:256-258.

Novitski, E. and L. Sandler. 1957. Are all products of spermatogenesis regularly functional? Proc. Nat. Acad. Sci. 43:318-324.

Provine, W. B. 1971. The origins of theoretical population genetics. The University of Chicago Press, Chicago and London.

Provine, W. B. 1994. The origin of Dobzhansky's genetics and the origin of species. In M.B. Adams (ed.), The evolution of Theodosius Dobzhansky. Princeton University Press, Princeton, NJ, pp. 99-114.

Renwick, H. H. 1971. The mapping of human chromosomes. Pages 81-120 in H. L. Roman, ed. Annual Review of Genetics, Annual Reviews, Inc., Palo Alto, Calif.

Ruder, A. 1985. Paternal age and birth-order effect on the human secondary sex ratio. Am. J. Hum. Genet. 37:362-372.

Sandler, L. and E. Novitski. 1957. Meiotic drive as an evolutionary force. Amer. Nat. XCI:105-110.

Sidirov, B. N., N. N. Socolov, and I. E. Trofimov, 1936. Crossing-over in heterozygoten Inversionen. I. Einfaches Crossing-over. Genetica 18:291-312.

Stern, C. 1936. Somatic crossing over and segregation in Drosophila melanogaster. Genetics 21:624-730.

Stern, C. 1949. Principles of Human Genetics. W. H. Freeman Co. San Francisco.

Stern, C. 1968. Genetic Mosaics and Other Essays. Harvard University Press, Cambridge.

Stern, C. and E. R. Sherwood. 1966. Gregor Mendel's letters to Carl Nägeli 1866-1873. In C. Stern and E. R. Sherwood, eds. The Origin of Genetics: A Mendel Source Book. W. H. Freeman and Co., San Francisco, Calif.

Stern, C. and E. R. Sherwood. 1966. The Origin of Genetics: A Mendel Source Book. Freeman, San Francisco.

Sturtevant, A. H. 1913. The linear arrangement of six sex-linked factors in Drosophila, as shown by their mode of association. Journal of Experimental Zoology 14:43-59.

Sturtevant, A. H. 1937. Autosomal lethals in wild populations of Drosophila pseudoobscura. Biol. Bull. Woods Hole 73:542-551.

Sturtevant, A. H. 1944. D. pseudoobscura. Ecology 25:476-477.

Sturtevant, A.H. 1946. Intersexes dependent on a maternal effect in hybrids between D. repleta and D. neorepleta. Proc. Nat. Acad. Sci. [Wash.] 32:84-87.

Sturtevant, A. H. 1965. A History of Genetics. Harper and Row, New York.

Sturtevant, A. H. and G. W Beadle. 1936. The relation of inversions in the X-chromosome of Drosophila melanogaster to crossing over and disjunction. Genetics 21:554-604.

Sturtevant, A. H. and G. W. Beadle. 1939. An introduction to genetics. W. B. Saunders, New York.

Sturtevant A. H. and Th. Dobzhansky. 1936. Inversions in the third chromosome of wild races of Drosophila pseudoobscura and their use in the study of the history of the species. Proc. Nat. Acad. Sci. [Wash.] 22:448-450.

Sturtevant A. H. and E. Novitski. 1941a. Sterility in crosses of geographical races of D. micromelanica. Proc. Nat. Acad. Sci. [Wash.] 27:392-394.

Sturtevant, A.H. and E. Novitski. 1941b. The homologies of the chromosome elements in the genus Drosophila. Genetics 26:517-541.

Szilard, L. 1960. Dependence of the sex ratio at birth on the age of the father. Nature 186:649-650.

Wallace, B. 1958. The comparison of observed and calculated zygotic distributions. Evolution 12:113-115.

Wright, S. 1941. On the probability of fixation of reciprocal translocations. Amer. Nat. 75:513-522.

Wright, S. 1966. Mendel's Ratios (in Stern and Sherwood 1966).

INDEX

ABOUT THE AUTHOR

Edward Novitski was born in Wilkes-Barre, Pennsylvania July 24, 1918. Graduated Elmer L. Meyers High School, 1936. Graduated Purdue University, B.Sc. with Distinction, 1938. Graduated California Institute of Technology, Ph.D., 1942. U.S. Army Air Force, 1942-1945. Fellow of John Simon Guggenheim Foundation, 1945-1946. Research Associate, University of Missouri, 1947-1948. Postdoctoral Fellow, California Institute of Technology, 1948-1950. Professor of Zoology, University of Missouri, 1951-1957. Head of the Biology Division, Oak Ridge National Laboratory, 1957-1958. Professor of Biology, University of Oregon, 1958 to present. Senior Postdoctoral Fellow of the National Science Foundation at the Commonwealth Scientific and Industrial Research Organization, Canberra, Australia, 1967. Fellow of John Simon Guggenheim Foundation, Fulbright Fellow, and National Science Foundation Fellow at the Division of Genetics, University of Leiden, The Netherlands, 1972-1974

www.ingramcontent.com/pod-product-compliance
Lightning Source LLC
Chambersburg PA
CBHW031831170526
45157CB00001B/262